**Books are to be returned on or before
the last date below.**

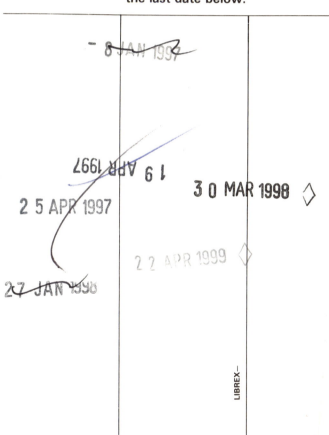

- 8 JAN 1997

19 APR 1997

2 5 APR 1997

3 0 MAR 1998

2 2 APR 1999

27 JAN 1998

LIBREX—

Large Concrete Buildings

Concrete Design and Construction Series

SERIES EDITORS

PROFESSOR F. K. KONG
Nanyang Technological University, Singapore

EMERITUS PROFESSOR R. H. EVANS† CBE
University of Leeds

OTHER TITLES IN THE SERIES

Concrete Radiation Shielding: Nuclear Physics, Concrete
Properties and Construction *by M. F. Kaplan*

Quality in Precast Concrete: Design—Production—Supervision
by John G. Richardson

Reinforced and Prestressed Masonry
edited by Arnold W. Hendry

Concrete Structures: Materials, Maintenance and Repair
by Denison Campbell-Allen and Harold Roper

Assessment and Renovation of Concrete Structures
by E. Kay

Concrete Bridges: Design and Construction
by A. C. Liebenberg

Fire Safety Design and Concrete *by T. Z. Harmathy*

Concrete Structures: Eurocode EC2 and BS 8110 Compared
edited by R. S. Narayanan

Concrete at High Temperatures
by Z. P. Bazant and M. F. Kaplan

Concrete Structures in Earthquake Region: Design and
Analysis *edited by E. Booth*

Fracture Mechanics and Structural Concrete
by B. L. Karihaloo

Fundamentals of High Strength High Performance Concrete
by E. G. Nawy

Large Concrete Buildings

Editors

B. V. Rangan

Professor and Head of the School of Civil Engineering
Curtin University of Technology, Western Australia

and

R. F. Warner

Professor of Civil Engineering
University of Adelaide, South Australia

Longman

Longman Group Limited
Longman House, Burnt Mill, Harlow
Essex CM20 2JE, England
and Associated Companies throughout the world

© Longman Group Limited 1996

First published 1996

British Library Cataloguing in Publication Data
A catalogue entry for this title is available from the British Library.

ISBN 0-582-10130-1

Library of Congress Cataloging-in-Publication Data
A catalog entry for this title is available from the Library of Congress.

Set by 21 in 10/12 Times New Roman

Printed and bound in Great Britain by Bookcraft (Bath) Ltd.

Contents

Contributors

Kim S Elliott

Kim S Elliott has been a lecturer in the Department of Civil Engineering since 1987. He previously spent seven years working in the precast concrete industry in the UK and overseas. He has authored two books on precast concrete framed buildings, and several papers reporting the research work carried out on connections in precast structures. He is a member of the FIP Commission on Prefabrication, and Chairman of a European Research Working Group on semi-rigid connections in concrete structures.

Contact: Department of Civil Engineering, University of Nottingham
University Park Nottingham NG7 2RD, United Kingdom
Phone: $+44$ 115 951 38 87; Fax: $+44$ 115 951 38 98
email: kim.elliott@nottingham.ac.uk

S K Ghosh

S K Ghosh is Director, Engineering Services, Codes and Standards, Portland Cement Association, Skokie, Illinois, and Adjunct Professor of Civil Engineering at the University of Illinois at Chicago. Dr Ghosh has co-authored one and edited three books dealing with the design, including earthquake resistant design, of concrete structures. He co-authored the chapter on earthquake engineering in two well-known structural engineering handbooks. He has published over one hundred and twenty papers on concrete-related subjects. He is currently a member of the American Concrete Institute Committee 318 Standard Building Code. He also currently chairs ACI Committee 435, Deflection of Concrete Structures.

Contact: Portland Cement Association, 5420 Old Orchard Rd,
Skokie, Illinois 60077-1083 USA
Phone: $+1$ 708 966 62 00; Fax: $+1$ 708 966 97 81

A W Hendry

A W Hendry is Professor Emeritus of Civil Engineering in the University of Edinburgh. He has been engaged in research on structural masonry for many years, and has been the author or co-author of four books as well as numerous papers on the subject. He is Past President and Honorary Member of the British

Masonry Society, whose journal 'Masonry International' he edits. He has served on many national and international committees concerned with codes of practice and has acted as consultant on projects relating to masonry buildings.

Contact: 145/6 Whitehouse Loan, Edinburgh EH9 2AN, Scotland
Tel: + 44 131 447 03 68; Fax: + 44 131 447 84 33

M C Griffith

M C Griffith is currently Head of the Department of Civil and Environmental Engineering and Senior Lecturer in Civil Engineering at the University of Adelaide. He has written numerous technical papers on the behaviour of buildings subject to earthquake ground motion and he has served on several national engineering code committees, in particular the Australian Earthquake Loading Code and the Elastomeric Structural Bearing Design Code. His principal research interests are in the areas of earthquake engineering and structural dynamics with particular reference to reinforced concrete and unreinforced masonry buildings.

Contact: Department of Civil and Environmental Engineering
The University of Adelaide, Adelaide 5005, Australia
Phone: + 61 8 303 51 35; Fax: + 61 8 303 43 59
email: mgriffith@civeng.adelaide.edu.au

Marita Kersken-Bradley

Marita Kersken-Bradley, a fire consultant in Munich, is involved with fire design for projects such as shopping malls, high-rise office buildings, industrial complexes and power plants. She is an appointed court expert and has published extensively in the fields of structural reliability and fire design. She was involved in the drafting of the Eurocodes, in particular the fire design requirements, and is a member of various national and international code committees and associations.

Contact: Kersken und Partner, Pienzenauer Strasse 18,
81679 München (80), Germany
Tel: + 49 89 988 039; Fax: + 49 89 9810 215

Karl Kordina

Karl Kordina is University Professor of Civil Engineering at the Technical University of Braunschweig and was Head of the Special Research Department 'Behaviour of Structural Members in Fire' in the period 1972–1987. He has authored more than 300 technical papers on research into concrete structures. His research interests include the non-linear analysis and design of concrete structures, under both normal temperatures and fire conditions.

Contact: Institut für Baustoffe, Massivbau und Brandschutz
Technische Universität Braunschweig, Beethovenstrasse 52,
D-38106 Braunschweig, Germany
Phone: + 49 531 391 54 11; Fax: + 49 531 391 45 73

B V Rangan

B V Rangan is currently Professor and Head of School of Civil Engineering at Curtin University of Technology, Perth, Western Australia. He has authored a text book and more than a hundred technical papers on research into concrete structures and concrete technology. His book *Reinforced Concrete* written with R F Warner and A S Hall will shortly go into its fourth edition. He has won several international awards and prizes for his research.

Contact: School of Civil Engineering
            Curtin University of Technology, GPO Box U1987
            Perth, Australia
            Phone: +61 9 351 70 48; Fax: +61 9 351 28 18
            email: rangan@macros.cage.curtin.edu.au

Karl-Heinz Reineck

Karl-Heinz Reineck is head of a research group at the Institute for Structural Design in the Civil Engineering Department of the University of Stuttgart, where he also teaches. His research interests include experimental and theoretical studies of shear, the dimensioning and detailing of structural concrete members, the design of shells and containment structures, as well as the computer-aided design of structural concrete. He has carried out expertises into the behaviour and design of critical regions of concrete offshore platforms, and numerous design checks of buildings and bridges. He is a member of ACI Commission 445 Shear and Torsion, FIP Commission III, and various Task Groups of the CEB.

Contact: Riegaläckerstrasse 3, Warmbronn, D-71229, Germany
            Phone: +49 49 7152 411 86; Fax: +49 7152 759 66

Jürgen Ruth

Jürgen Ruth is currently a member of BGS Ingenieursozietat – Consulting Civil Engineers – Frankfurt. He is involved with the design of large buildings, and in particular with the optimization of structural systems. Beside these activities he teaches design of reinforced concrete and structural analysis at the Polytechnic, Frankfurt. He was previously a member of the Institute for Structural Design at the University of Stuttgart, where he undertook doctoral research into friction in joints in structures, under the supervision of Professors Schlaich and Schäfer. The results of this research have been published in his thesis and in several technical papers and colloquium reports.

Contact: Mozartring 39 D-63543, Neuberg, Germany
            Phone: +49 6183 728 92; Fax: +49 6183 72 8 92

George Somerville

George Somerville is Director of Engineering at the British Cement Association and a Visiting Professor at Imperial College, London, and at Kingston University. His research interests have included precast and composite construction, bridge design, structural detailing and model testing; the current dominant interest is durability and design life, both for new construction and existing structures. In all, Dr Somerville has co-authored three books, and nearly 100 papers on all aspects of the behaviour of structural concrete. He has been actively involved in the production of UK and European codes of practice, and in the activities of international bodies such as CEB, FIP, and IABSE.

Contact: British Cement Association
            Century House, Telford Avenue
            Crowthorne, Berkshire RG45 6YS, England
            Phone: +44 1344 76 26 76; Fax: +44 1344 76 12 14

R F Warner

R F Warner is Professor of Civil Engineering at the University of Adelaide. He has co-authored text books on reinforced concrete, prestressed concrete and engineering planning and design, as well as numerous technical papers dealing with research into concrete structures, and engineering education. His current research interests include the non-linear analysis and design of concrete structures, and the pathology, assessment and repair of concrete buildings.

Contact: Department of Civil and Environmental Engineering
            The University of Adelaide, Adelaide 5005, Australia
            Phone: +61 8 303 54 51; Fax: +61 8 303 43 59
            rwarner@civeng.adelaide.edu.au

Preface

This book is intended for practising structural engineers who are engaged in the planning and design of concrete buildings, and who already have a basic understanding of structural concrete behaviour and of the concepts of structural analysis and design.

In order to undertake the design of a concrete building, the structural engineer must have a good understanding of the way such buildings behave under normal in-service conditions and at high overload, and must also be familiar with the calculation procedures, based on limit states concepts, which are used for checking the design requirements of good serviceability and adequate strength. Detailed information on the behaviour of concrete structures and on relevant methods of analysis and design can be readily found in a wide range of existing text books which have been prepared for undergraduate and postgraduate courses on structural analysis and design.

However, in order to ensure that the building is economically, functionally and structurally successful, the designer must address a variety of additional important questions. These vary in nature, depending on the specific features of the building. For example, special structural questions can arise, such as the need or otherwise for articulation in a large building, and hence for structural joints. Another example is the problem in a tall building of differential shortening of the columns. Planning matters such as choosing an appropriate design life or determining the relative costs of initial construction and ongoing maintenance always need careful consideration, preferably in the early stages of the project. Other considerations, for example the provision of adequate fire resistance, require attention in both the initial planning stage and in the detailed stage of structural design.

This book provides information on a range of important topics which arise repeatedly in the planning and design of concrete buildings. The topics dealt with include: design-life performance, durability and life-cycle planning; structural modelling and approximate methods of analysis; computer-aided analysis and design; evaluation and accommodation of column-length changes in tall buildings; the use of strut-and-tie modelling for the rational detailing and design of complex regions of concrete buildings; planning and structural design for fire resistance; the applications of high-strength concrete in buildings; the uses of masonry in concrete buildings; precast concrete skeletal construction; and the need, or otherwise, for movement joints in large buildings. In each case we have sought out experts of international standing to provide a thorough, state-of-the-art treatment of these topics.

We are well aware that the coverage is not extensive or complete. Given the space limitations which inevitably apply in the preparation of a text such as this, we have concentrated on a relatively small number of important topics which are of continuing importance in the design of concrete buildings. Each chapter provides a general introduction to the relevant concepts, and gives practical details for use by the designer. Where appropriate, additional information can be obtained from the useful references provided.

R F Warner, Adelaide   B V Rangan, Perth
December, 1995

1 Life Cycle Planning: Durability, Performance and Design Life

G. Somerville

This chapter recognizes that the performance requirements for structures have changed, with greater emphasis on performance in service with time. Durability has become much more significant, where lack of technical performance must not interfere with the maintenance of function at minimum cost (whole life costing). At the same time materials, and design and construction practices have also changed.

This chapter begins by summarizing these changes, before outlining current thinking in terms of developing service life concepts as an essential part of the initial design process. The approach has to be holistic, in integrating material, design and construction issues, to achieve greater control over durability. However, it is first essential to have a clear statement of the required performance criteria; this can only come from a whole life costing approach.

1.1 Introduction

Initial design and construction costs are a fraction of the total cost of the investment represented by any structure. Costs-in-use are significant in themselves, and any unnecessary additions to these, due to a failure in technical performance in service, are not welcome. This is illustrated graphically in Fig. 1.1, using the iceberg analogy usually attributed to Blanchard;[1.1] initial design can have an influence on all the segments below the waterline in this figure, but especially on maintenance, energy and operating costs.

Awareness of these issues has been high for some considerable time. However, detailed knowledge and routine mechanisms for every day use have been lacking, to permit design and planning on a life cycle basis. Owners are nevertheless becoming more demanding in this regard; for designers, there is then a need to go beyond simple conceptual design and numerical calculations, to consider the role of the structure as an integral part of the complete artefact – and its contribution, over a period, to the financial and functional well-being of that artefact.

We are therefore at a period of change, in terms of owners' perceived needs and performance requirements. Also we can now detect trends and lessons from in-service feedback over a period of 40–50 years of concrete construction. During that same period, there has been considerable change in materials, in design standards and construction methods. All of these issues will be reviewed in this chapter, prior to outlining a design approach with a strong emphasis on technical performance with time, commensurate with the perceived functional and financial needs of owners.

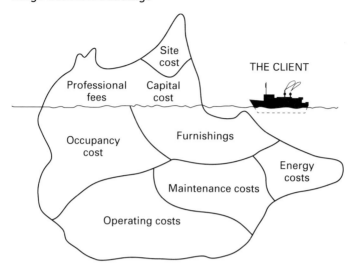

Fig. 1.1 Total life cycle costs

1.2 Trends in Construction and Performance Requirements

1.2.1 Materials

Materials have become stronger. Figure 1.2 shows the increase in cement strength this century, as measured on standard concrete cubes with a water/cement ratio of 0.6. With current materials, it is possible to have much higher strengths by varying the mix ingredients or proportions, or both. It has been demonstrated in a number of countries that this can be done routinely using either precast or ready-mixed concrete up to strengths of 80–100 N mm^{-2}, or more. An important economic factor is that increase in strength is not proportional to the increase in cost, as shown typically in Fig. 1.3.

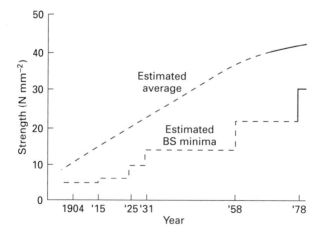

Fig. 1.2 Increase in cement strengths

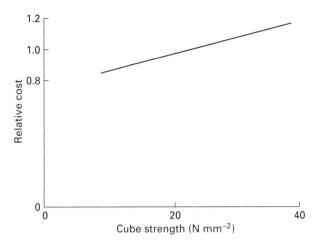

Fig. 1.3 Concrete strength versus cost

A similar picture emerges with reinforcement and prestressing steel. Not only has strength increased, but an increasing proportion of that strength is permitted in design. High strength reinforcement is only marginally more expensive than mild steel: 50 years ago mild steel was the norm; now, the usage of high strength bars is about 90 per cent.

Other notable trends include:

- the use of admixtures and additives to modify the properties of either the fresh or hardened concrete;
- the introduction of coatings or other protective systems, to give increased corrosion resistance;
- the development of prestressing systems to provide high concentrated forces.

These trends have generally been in response to changes in performance, design or construction needs. Nothing ever stands still; these trends, or others like them, will continue into the future.

1.2.2   Design

Change in design methods or standards are most easily detected by looking at codes of practice over the years. In terms of methods, permissible stress design was first replaced by the load factor method, followed more recently in many countries by limit state design. Increased research and development, together with the growing use of computers, has led to the parallel development of sophisticated analytical and design techniques. These trends, together with changes in materials, have led to increased strength, lower load factors and higher permissible stresses – all leading to lighter and more flexible structures.

1.2.3   Construction

In the first half of this century, concrete was mixed and placed on-site. That practice has changed significantly. Typically, more than 70 per cent of concrete used in construction is now mixed off-site and delivered by truck, the bulk of it being covered by a quality assurance scheme. Additionally, it is now possible to obtain a wide range of structural precast components, made under factory conditions, with good quality control.

More generally – and driven by a demand to reduce construction time, requiring more attention to buildability – construction methods and economics have changed significantly in recent years. Project management now looms large on any scheme, and developments in temporary works (e.g. sliding and flying formwork) have had to keep in step with those for permanent structures.[1,2] Prestressing also plays a larger role in buildings, in responding to demands for bigger spans and the better use of space.

All of this has radically changed the pecking order in floor systems, with flat slabs now more common, often incorporating shearheads over columns, and frequently prestressed. Moreover, the whole construction process has been streamlined – 'simplify', 'standardize' and 'repetition' have all become buzz words – and even the make-up of the industry itself has been changed to meet these new demands. Typically, the cost of a modern concrete frame is as shown in Table 1.1, with specialist subcontractors concentrating on each aspect.

1.2.4   Buildings – Lessons from Past Performance in Service

The general material, design and construction factors mentioned above have also had a direct effect on buildings. Buildings are now lighter and more flexible, and possibly more sensitive or vulnerable to changes in design/construction practice, or to misuse, or accident in service. Figure 1.4, based on data from Sweden, illustrates this trend, by showing how building weights have almost halved over a period of 50 years.

However, there are other factors at work. Table 1.2 compares the relative construction costs for different parts of similar 10-storey office buildings designed in the UK in 1960 and 1985. The most significant changes, over this 25-year period, are the rise in the cost of services and cladding, and the decrease in that for the structure including foundations. Owners and occupiers now have different needs, and expect higher standards in furniture and fittings.

Table 1.1 Approximate cost breakdown of a concrete frame (1989)

Construction component	Percentage of total cost
Concrete	14–16
Pumping/placing	11–13
Reinforcement	25–30
Formwork	45–50

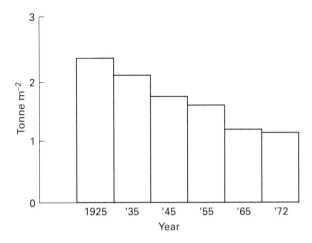

Fig. 1.4 Decrease in building weights (Swedish Survey, 1981)

Taken to its logical conclusion, the trend in Table 1.2 would see the basic structure provided more-or-less free by the year 2010! It is unlikely that the Table 1.2 trend will continue, however. Very recently, environmental issues have intervened, coupled with more consideration of the needs of occupants. Energy conservation is an important factor, and designs now aim to minimize the extent of artificial services, with potential health benefits. By fully utilizing concrete's thermal capacity, the problem of summer overheating (now exacerbated by IT heat loads) can be reduced, minimizing or even eliminating the need for air-conditioning. Thermal comfort is an important parameter, in both hot and cold conditions.

Since this chapter is concerned predominantly with durability and design life, no review of past performance would be complete without considering defects, in terms of their nature, where they occur and their underlying causes. Lack of durability, in these terms, is what increases the size of the maintenance segment in Fig. 1.1.

There have been numerous surveys of defects over the years. With minor variations, most tell the same story. That story can be summarized by reference to Paterson,[1.3] who reported on a survey of 10 000 defects in France, covering a range of materials and building types.

Table 1.2 Comparative breakdown of construction costs of a 10-storey office block in 1960 and 1985

	Percentage of total cost	
	1960	1985
Foundations	4.9	2.9
Superstructure	20.3	8.6
Cladding	16.8	28.6
Finishes	24.8	9.2
Services	33.2	50.7

Firstly, Paterson identified where the defects occurred; this is summarized in Table 1.3. In fact, slightly over half of the defects occurred in the external envelope, with only 25 per cent in the basic structure – and a large proportion of these were in the foundations. Paterson next analysed the cause of the defects, and this is given in Table 1.4; approximately half were attributed to construction, which inevitably means poor standards of workmanship – due either to a lack of skill or understanding, or to poor communications or quality control.

However, the 37 per cent attributed to design is also of concern. A breakdown of this is given in Table 1.5. The overwhelming cause is poor detailing. If this is related to Table 1.3, then detailing of the building envelope becomes significant. This will be touched on later in this chapter, since if there are faults in the overall envelope, leading, say, to the ingress of water, then this can have a subsequent influence on the durability of the structure itself. Taken together, however, Tables 1.3–1.5 demonstrate that lack of performance in durability terms has little to do with conventional design calculations, and everything to do with an overall understanding (both in concept and detail) coupled with the quality of construction.

Table 1.3 Percentage of defects in different parts of buildings (after Paterson,[1.3] sample of 10 000 defects)

Part of building	Percentage of defects
External envelope, including masonry	27
Cladding and roofing	25
Basic structure, including foundations	25
Internal fittings, services etc.	23

Table 1.4 Causes of defects in buildings (after Paterson[1.3])

Cause	Percentage of defects
Design	37
Construction	51
Faulty materials	4.5
Faulty maintenance	7.5

Table 1.5 Breakdown of design faults in Table 1.4 (after Paterson[1.3])

Fault	Percentage of defects
Detailing	78
Defective overall concept	14
Use of unsuitable materials	5
Error in calculations	3

1.2.5   Buildings – Summary of Trends in Performance Requirements

In general, owners are becoming much more aware of the time factor, in performance terms, and increasingly expect their designers to take this into account. This manifests itself at two levels:

- the time factor, during design and construction, which puts a strong emphasis on speed of construction, buildability and project management;
- the time factor for the building in service, with a greater emphasis on management and maintenance. This has also to be considered at the design stage, since it is integrated both with the timing of the projected financial outlay, and with future functional needs, including possible changes in use or refurbishment.

In addition to the time factor, there are other issues of growing importance which can influence the planning of service life performance. A list of such factors might include:

- the provision of more usable space, with a capability for modification;
- environmental or health issues;
- the integration of the structure both with the cladding and the services;
- the provision made for exceptional loads, including fire, impact or explosions.

1.3   Service Life Concepts in Design

1.3.1   Introduction

A thorough review of the practicalities of design life concepts is contained in the proceedings of a colloquium organized by the British Group of IABSE.[1.4] The subject was durability and the context was a perception of lack of performance in service of structures built during the previous 40 years measured against extensive research and development work on deterioration mechanisms and their effects. Six key questions were addressed as follows.

1. What do we understand by the term/concept of design life?
2. Can/should we, consciously design, detail and build structures for a specified (albeit 'notional') design life?
3. How do we develop the concept in practical terms? What should the design life be for different categories of structure or component?
4. Do we have the knowledge, and the design and construction techniques, to tackle this at this time?
5. What are the real factors involved – practical and technical? What are the financial implications? What are the advantages – why should we do it?
6. Do we really need to develop these concepts at all in order to do 'better', or does the answer lie in improving existing methods, developing greater awareness, stepping up training and making a greater commitment to quality?

One school of thought was that durability would be improved remarkably if we actually managed to achieve on site what we set out to do on the drawing board; there is some truth in this, as Tables 1.3–1.5 indicate.

A second school of thought was that there were too many unknowns at the design stage to permit a quantitative approach to design life. What was required for greater durability, therefore, was a rationale, which included the following features.

1. Systematic design to cope with known hazards, considering both risk and consequence (possibly involving a design review/audit of durability, linked to the importance or criticality of the structure of structural element).
2. Relative insensitivity to:

    - marginal departures from the design assumptions;
    - local defects or movements;
    - environmental change (both micro- and macroclimate).

3. Buildability, and not total dependence on perfect workmanship and compliance with specifications. The establishment of minimum standards of workmanship and relevant specifications, together with appropriate (quantifiable) methods for checking that they are met.
4. Provision of good access for all items requiring inspection or maintenance.
5. Incorporation of early warning visible signs of serious defects.
6. Limitation of movement, during the expected useful life, so that the function of the structure is unimpaired.
7. Capability to allow some change in use.

This is an attractive scenario, based on existing knowledge and practice. However, it was felt at the colloquium that the existence of a target life would be useful in a number of ways, including:

- the development of material/component specifications;
- assessing alternative design strategies in comparative as well as absolute terms;
- in the short term, as a technique to develop balanced approaches to design based on lifetime concepts;
- the introduction of the concepts of replaceability, maintainability and consideration of inaccessible parts;
- use of 'yardsticks of quality' along the lines of agrément assessments;
- formalizing already established techniques, e.g. the use of accelerated testing in establishing 'life' under fatigue loading.

In summary, the development of modelling of deterioration processes is proceeding quickly, nearing the stage where it can be incorporated into a design system. This enables technical life to be assessed with some precision, and matched against future projections for finance and function. A key argument is that greater durability is not necessarily the desired end result; what is needed is more *control*, in predicting technical life, commensurate with financial outlay and functional obsolescence. This is perhaps most clearly seen in the UK in the case of bridges; for primary routes, bridges designed for a notional life of 120 years have become obsolete within 30 years, due to increases in traffic and in axle loads.

1.3.2   Current Developments

The key first step is to define a target service life. One approach is to do this in a simple qualitative way. This is illustrated in Fig. 1.5, based on work within CEB, the European Committee for Concrete. Part of the argument is that no greater precision is either necessary or possible, as far as the base structure is concerned.

However, as may be seen from Table 1.2 and Fig. 1.4, the structure is becoming less significant in terms of the totality of the building; there is therefore an argument for moving towards the type of approach which is already prevalent for services and cladding – the conscious recognition of maintenance as a key item, plus the need for replacement.

The trend towards a quantified approach can be illustrated by reference to a recent British Standard.[1.5] There, categories of design life for buildings, are as shown in Table 1.6. This is supported by categories for components as given in Table 1.7, and by a definition of maintenance levels, Table 1.8.

It must be stressed at this point that the figures in Table 1.6 represent a target to be aimed at, *as part of the design concept*, in assessing alternative options and strategies for achieving a level of technical performance, which is compatible with the client's future functional and financial requirements (as perceived at the design stage). They represent design criteria (analogous to setting limits on crack widths or deflections) and in no way should be taken as a definition of actual life – that merely confuses the issue, and raises the spectre of liability in many people's minds. Actual (useful) life will depend on the owner's future financial and functional planning – and on the actions he takes in terms of maintenance and upgrading. These actions do not invalidate the argument for a target life as part of the design concept.

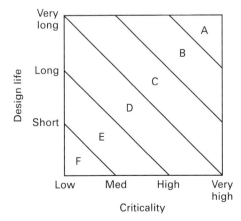

Fig. 1.5 CEB qualitative approach for measuring the significance of design life

Table 1.6 Categories of design life for buildings[1.5]

Category	Description	Building life for category	Examples
1	Temporary	Agreed period up to 10 years	Non-permanent site huts and temporary exhibition buildings
2	Short life	Minimum period 10 years	Temporary classrooms; buildings for short-life industrial processes; office internal refurbishment, retail and warehouse buildings (see Note 1)
3	Medium life	Minimum period 30 years	Most industrial buildings; housing refurbishment
4	Normal life	Minimum period 60 years	New health and educational buildings; new housing and high quality refurbishment of public buildings
5	Long life	Minimum period 120 years	Civic and other high quality buildings

Note 1: specific periods may be determined for particular buildings in any of categories 2 to 5, provided they do not exceed the period suggested for the next category below in the table; for example many retail and warehouse buildings are designed to have a service life of 20 years.
Note 2: buildings may include replaceable and maintainable components.

Table 1.7 Categories of design life for components or assemblies[1.5]

Category	Description	Life	Typical examples
1	Replaceable	Shorter life than the building life and replacement can be envisaged at design stage	Most floor finishes and service installation components
2	Maintainable	Will last, with periodic treatment, for the life of the building	Most external claddings, doors and windows
3	Lifelong	Will last for the life of the building	Foundations and main structural elements

Table 1.8 Maintenance levels (other than daily and routine cleaning)[1.5]

Level	Description	Scope	Examples
1	Repair only	Maintenance restricted to restoring items to their original functions after a failure	Replacement of jammed valves; reglazing of broken windows
2	Scheduled maintenance plus repair	Maintenance work carried out to a predetermined interval of time, number of operations, regular cycles etc.*	Five-yearly external joinery painting cycle. Five-yearly re-coating of roof membrane with solar reflective paint
3	Condition-based maintenance plus repair	Maintenance carried out as a result of knowledge of an item's condition (the condition having been reported through a systematic inspection (procedure))	Five-yearly inspections of historical churches etc. leading to planned maintenance

* The length of the regular maintenance cycle is an important factor. It may be stated in the brief or agreed while the design is being developed.

BS 8210 recommends systematic inspections as follows:
(a) continuous, regular observation by the building user as part of the occupancy of the building;
(b) annual visit inspection of the main elements under supervision of suitably qualified personnel; and
(c) full inspection of the building fabric by suitably qualified personnel at least every five years.

There may be statutory requirements for more frequent inspections.

System	Criticality*	Target life before replacement						Capital cost plan £1	Cost in use target x years £K/annum
		>5	5–10	10–20	20–40	40–100	<100		
Foundations	A	▮	▮	▮	▮	▮	▮		
Structure	A	▮	▮	▮	▮	▮	▮		
External walls	A	▮	▮	▮	▮	▮	▮		
External cladding	B	▮	▮	▮	▮				
Curtain walling	B	▮	▮	▮	▮				
Windows	A	▮	▮	▮	▮				
Roof covering	B	▮	▮	▮					
Rainwater goods	B	▮	▮	▮					
Internal partitions	B	▮	▮	▮					
Demountable partitions	C	▮	▮	▮	▮				
Doors and ironmongery	C	▮	▮	▮					
Finishes generally	A	▮	▮	▮					
Raised floor	B	▮	▮	▮	▮				
Floor coverings	B	▮	▮						
Suspended ceilings	C	▮	▮	▮					
Fittings and furnishings	B	▮	▮						
HVAC plant	B	▮	▮	▮					
HAVC ducts, pipes etc.	B	▮	▮	▮					
Water installations	B	▮	▮	▮					
Public health services	B	▮	▮	▮	▮				
Drainage	B	▮	▮	▮	▮	▮			
Sanitary fittings	A	▮	▮	▮	▮				
Electrical plant, switchgear etc.	A	▮	▮	▮	▮				
Electrical installations	B	▮	▮	▮					
Luminaire	C	▮	▮						
Internal decorations	C	▮							
External decorations		▮						(say) 100	(say) 1500
					Capital cost budget (£k)				
				Total cost in use target over x years (£pv)					

* The question of criticality can be addressed by a simple grading: A–highly critical, failure causing cessation of operation and disruption during remedial work; B–critical, lowering working efficiency, remedial work out of normal hours; C–not critical, requiring remedial work but not immediately essential. Other definitions may be developed, but these are given to illustrate the performance profile.

Fig. 1.6 Outline building performance profile[1.6]

It is essential that Table 1.6 is supported by Tables 1.7 and 1.8. A maintenance strategy is essential (Table 1.8). However, it is also necessary to consider parts of the building, including the structure, in terms of Table 1.7. For example, if reinforcement corrosion is a design issue, and if a particular structural element is difficult to maintain or replace, then more stringent design requirements might be introduced (e.g. thicker concrete cover, or protective systems).

The Table 1.6–1.8 approach can be developed further, as illustrated by White.[1.6] This is based on performance profiles, as illustrated in Fig. 1.6. The suggestion is that such profiles should be the basis of a dialogue between client and designer at an early stage, as an aid to defining performance and cost-in-use requirements, while simultaneously identifying the most economic cycles for repair and/or replacement.

In the example shown, the target life for the structure is very much 'long life'. However, as the concept is developed, that could change – thus opening up the possibility of major upgrading or redevelopment, should the expectations of the occupier change, or, say, should a significant change in land values make an increase in prime above ground space attractive. These possibilities could be programmed to coincide with scheduled repair or replacement cycles for services, cladding etc.

All that is being suggested here is that the management of buildings in service, in terms of function and finance, should be considered at the design stage. An important aspect is that the technical performance of the structure should not interfere unduly with that management – and preferably be integrated with it. This has a strong influence in designing for compatible durability – the issue which is considered next.

1.4   Durability

In 1986, Somerville[1.7] drew a distinction between:

- the specification, production and placing of durable concrete;
- the design and detailing of structures that will be durable.

That distinction will be maintained here.

1.4.1   Durable Concrete

The specification of durable concrete is done in codes of practice, backed by supporting standards on materials, components and testing. Each code is different in detail – sometimes significantly so – but the basic features usually are:

(a)   a classification of exposure conditions;
(b)   an emphasis on achieving low permeability, translated in practice into detailed recommendations on mix constituents and proportions, cover, compaction and curing;
(c)   specific recommendations regarding certain forms of aggressive action, e.g. sulphate attack, abrasion, freeze–thaw, alkali–silica reaction.

(a) and (b) together provide a minimum general requirement, which, when associated with specified covers, is expected to deal with corrosion resistance.

The approach may be illustrated by reference to the British Code, BS 8110.[1.8] Table 1.9 shows the definition of exposure conditions. Table 1.10 indicates both concrete cover and mix proportions for each exposure condition. The fact that this is intended to deal with corrosion may be seen by comparing Table 1.10 with Table 1.11 (the corresponding requirements for plain concrete).

An examination of other national codes in Europe, Australia and America shows the same basic approach, but with differences in detail and in emphasis. For example, water/cement ratio is seen as the key durability parameter in continental Europe, whereas concrete grade is considered to be a better parameter in the UK, Australia and America, partly for ease of compliance checking and partly to cover a wide range of cement types. Supporting material standards are

Table 1.9 Exposure conditions (BS 8110)[1.8] (reproduced by kind permission of the British Standards Institution)

Environment	Exposure conditions
Mild	Concrete surfaces protected against weather or aggressive conditions
Moderate	Concrete surfaces sheltered from severe rain or freezing whilst wet. Concrete subject to condensation. Concrete surfaces continuously under water. Concrete in contact with non-aggressive soil
Severe	Concrete surfaces exposed to severe rain, alternate wetting and drying or occasional freezing or severe condensation
Very severe	Concrete surfaces exposed to sea water spray, de-icing salts (directly or indirectly), corrosive fumes or severe freezing conditions whilst wet
Extreme	Concrete surfaces exposed to abrasive action (e.g. sea water carrying solids, flowing water with pH $\leqslant$ 4.5, machinery or vehicles)

Table 1.10 Nominal cover to all reinforcement (including links) to meet durability requirements (BS 8110)[1.8] (reproduced by kind permission of the British Standards Institution)

Conditions of exposure (see Table 1.9)	Nominal cover				
	mm	mm	mm	mm	mm
Mild	25	20	20*	20*	20*
Moderate	–	35	30	25	20
Severe	–	–	40	30	25
Very severe	–	–	50†	40†	30
Extreme	–	–	–	60†	50
Maximum free water/cement ratio	0.65	0.60	0.55	0.50	0.45
Minimum cement content (kg m^{-3})	275	300	325	350	400
Lowest grade of concrete	C30	C35	C40	C45	C50

* These covers may be reduced to 15 mm provided that the nominal maximum size of aggregate does not exceed 15 mm.
† Where concrete is subject to freezing whilst wet, air-entrainment should be used (see 3.3.4.2[1.8]).
Note 1: this table relates to normal-weight aggregate of 20 mm nominal maximum size.
Note 2: for concrete used in foundations to low rise construction (see 6.2.4.1[1.8]).

Table 1.11 Durability of unreinforced concrete made with normal-weight aggregates of 20 mm nominal maximum size (BS 8110)[1.8] (reproduced by kind permission of the British Standards Institution)

Conditions of exposure (see Table 1.9)	Concrete not containing embedded material		
	Maximum free water/cement ratio	Minimum cement† content (kg m^{-3})	Lowest grade of concrete
Mild	0.80	180	C20
Moderate*	0.65	275	C30
Severe	0.60	300	C35
Very severe	0.55	325	C35‡
Extreme	0.50	350	C45

* See 6.2.4.1[1.8] for concrete used in foundations to low rise construction.
† Inclusive of g.g.b.f.s. or p.f.a. content.
‡ Applicable only to air-entrained concrete.
Note 1: see 6.2.3.2[1.8] for freezing and thawing conditions and 6.2.4.3[1.8] for adjustments to the mix proportions.
Note 2: the lowest grades of concrete may be reduced by not more than 5 N mm^{-2}, provided there is evidence showing that with the materials to be used, this lower grade will ensure compliance with the required minimum cement content and maximum free water/cement ratio.

generally consistent on matters of detail (e.g. chloride and sulphate contents, aggregates, admixtures).

The essential weakness in the system, as far as corrosion resistance is concerned, is that the definition of exposure conditions is too general, and does not reflect the critical microclimate conditions which can occur locally on the surface of concrete elements.

For specific aggressive actions that predominantly affect the concrete, durability design solutions again appear in specification form. This is illustrated in Table 1.12, where for four such actions, the general approach is outlined, together with some brief comments on each action. By and large, this system works well.

Table 1.12 Types of aggressive action for which material specifications have been developed

Aggressive action	General approach	Comments
Sulphate attack	Quantify the action	Specific material and mix proportions are recommended in most codes for defined ranges of sulphate concentration.
Alkali–silica reaction		The basic reaction and its possible effects are now well understood. Recommendations to minimize the risk of damage are published.
Freezing and thawing	Define ranges of intensity for it	Dealt with by choice of materials, mix proportions and concrete grade. Air entrainment for lower grades. Detail to minimize exposure to moisture
Abrasion	Produce a specification for each range	Specifications to cover aggregate properties, concrete grade and mix proportions, compaction and curing, methods of finishing etc.

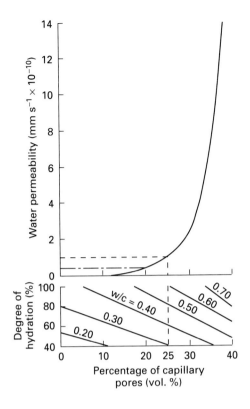

Fig. 1.7 Typical influence of w/c ratio on permeability

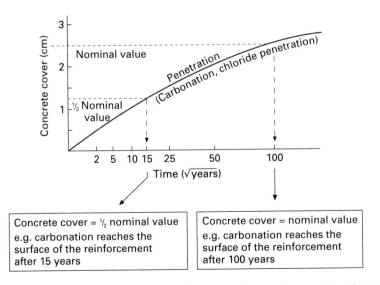

| Concrete cover = ½ nominal value e.g. carbonation reaches the surface of the reinforcement after 15 years | Concrete cover = nominal value e.g. carbonation reaches the surface of the reinforcement after 100 years |

Fig. 1.8 Importance of achieving the specified cover, with respect to corrosion resistance

1.4.2   Reinforcement Corrosion

Corrosion is arguably the most serious durability issue for structural concrete, principally due to chlorides arising from the use of de-icing salts on roads, but also coming from sea water. Corrosion can also occur, in the presence of oxygen and water, due to the carbonation of concrete; while a less serious threat than that due to chlorides, it is probably the greatest risk for buildings located away from a salt-water environment.

Much is now known about the mechanisms of corrosion. In Europe, activity has centred on the CEB.[1.9,1.10,1.11] Figure 1.7, taken from Ref. 1.11, demonstrates the influence of water/cement ratio (w/c) on permeability. Figure 1.8, from the same source, clearly indicates the importance of achieving the specified cover (a key workmanship factor – see Table 1.4). Additionally, the most critical moisture conditions have been established, as indicated in Table 1.13. However, this table also indicates a specification problem for durability generally, since the critical relative humidity is different for separate deterioration processes.

In attempting to go beyond the simple specification typified by Tables 1.9–1.11, current efforts at a more rigorous design approach are directed at the following issues.

1. A better definition of exposure conditions, directly aimed at corrosion, and with a strong emphasis on microclimate adjacent to, or physically in, the concrete cover to the reinforcement. To do this properly, distinctions have to be made between carbonation and chloride-induced corrosion – and even between sea water and de-icing salts. Such an approach has been developed for draft CEN standards for concrete and structural design.

2. The development of the simple two-phase mechanism illustrated in Fig. 1.9. Clearly the gradient of the t_i line has to be established, as well as the 'significant level of damage' on the vertical axis. Establishing a limit for the t_0 phase is not straightforward either, when chlorides are involved, since corrosion initiation depends not only on the permeability of the concrete cover, but also on the build-up of a critical threshold level of chlorides (which is dependent on cement chemistry among other things).

Table 1.13 Significance of moisture state in influencing different durability processes

Effective relative humidity	Process				
	Carbonation	Corrosion of steel in concrete which is		Frost attack	Chemical attack
		carbonated	chloride contaminated		
Very low ($<45\%$)	1	0	0	0	0
Low (45–65%)	3	1	1	0	0
Medium (65–85%)	2	3	3	0	0
High (85–98%)	1	2	3	2	1
Saturated ($>98\%$)	0	1	1	3	3

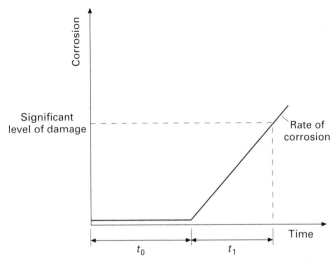

Fig. 1.9 Two-phase mechanism for corrosion of reinforcement. t_0 is the time for the aggressive agent to reach and activate the reinforcement; t, is the time for the corrosion to reach a level that causes significant damage

In translating this approach into a practical design method, it will be necessary to assemble the following elements.

1. A definition of 'loads' (aggressive actions) in both nature and intensity.
2. Acceptable performance criteria, i.e. a target service life, plus associated criteria defining when performance is considered to be unsatisfactory.
3. Predictive models, which quantify the effects of the aggressive actions and relate those predictions to the defined performance criteria.
4. Margins/factors of safety, introduced to take account of variability in the system and inaccuracy in the models, to give an acceptable probability that the required performance is achieved for the target service life.

These four basic elements have to be supported by:

*   material specifications, i.e. essentially for the required concrete quality, but also for concrete cover where corrosion is involved;
*   workmanship standards, e.g. tolerances on concrete cover;
*   a maintenance (and upgrading) strategy, perhaps based on the performance profile approach (see Fig. 1.6).

Clearly, we are still some way from having such a design method in place. Nevertheless, the data generated to date are of enormous benefit in considering the next step in this chapter – the design and detailing of buildings that will be adequately durable.

1.4.3.   Buildings – a Practical Design Approach to Durability Design

Environmental-Exposure Conditions
A first, essential step is to establish the general exposure conditions, and the possible existence of particularly aggressive actions. A starting point here is the

local equivalent of Table 1.9, but being mindful of the potentially critical microclimate as typified in Table 1.13. Clearly, in individual cases, the location and local topography will be important – as will the influence of wind, water and temperature. How rigorous the designer need be will depend on a number of factors, including:

- the client's specified requirements, in terms of function, appearance and maintenance;
- the nature of the building, its location and its importance;
- the performance of existing structures in the same general area;
- the overall design concept and the relationship of the structure to the cladding;
- the materials to be used, the quality of construction and the design solution to be adopted.

Performance Requirements

Initially, this takes us into the realm of Tables 1.6 and 1.7, in relation to the structure as a whole and to its component parts. Linked to that is Table 1.8 on maintenance levels. Choices have to be made, and Fig. 1.7 is one systematic way of doing that. For the structural elements, this will give some indication of what is required in time terms, before then delving into detailed aspects, such as those in Tables 1.10–1.12. In some circumstances, it may be prudent to go further, as far as the long-life structural elements are concerned. In designing these – both for strength and serviceability – conventional factors of safety will be used. Should these be eroded, due to deterioration, then what is the maximum acceptable technical performance – and can this be maintained via the proposed lifetime plan for the structure? The principle of this is illustrated in Fig. 1.10. This is a common issue in the assessment of existing structures; it perhaps requires more consideration in new construction, in evaluating the relative merits of different options in performance profile terms.

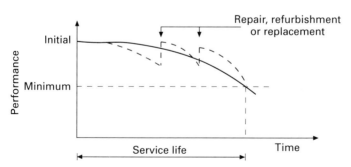

Fig. 1.10 The principle of checking that reduction in structural performance does not reach a critical minimum level

1.4.4   Buildings – Detailing for Durability

With many modern buildings, the structure is virtually totally enclosed within the cladding (which often has a projected shorter life (see Fig. 1.7)). With others, parts of the structure may be expressed architecturally, with direct exposure to the elements. Plainly, this will define the basic exposure conditions (e.g. Table 1.9), and hence the materials to be used.

The structural concrete used in many buildings will therefore be of the 'indoor' type, with low relative humidities and perhaps vulnerable to carbonation, but not necessarily corrosion due to carbonation (see Table 1.13). It is then necessary to ensure that the assumed exposure condition does not change in service to something that is much more critical. This is most likely to happen round the perimeter of the building due to deterioration in the façade. This is very much influenced by detailing, and there are two particular issues that require consideration – allowing for movement and dealing with water.

Allowing for Movement

Movement can arise in a number of different ways and, if not allowed for or controlled, can lead to the ingress of aggressive actions and hence a change in the exposure conditions assumed in design; if this happens anywhere in the façade (irrespective of the material used) then it may increase the risk of corrosion. Movement may also lead to restraints or deformations which could make the building more vulnerable in durability terms.

For structural concrete perhaps the most significant form of deformation in this context is cracking. Cracking can occur for very many reasons, some of these are listed in Fig. 1.11, taken from Reference 1.11. These mechanisms can be considered in two separate categories.

1.   Those due to movements generated within the concrete (surrounded by a box in Fig. 1.11) – *intrinsic cracks*.
2.   Those due to externally imposed conditions – *extrinsic cracks*.

Intrinsic cracks are due to the presence of restraints, either within the concrete element itself or due to boundary conditions. Their possible significance may be due to the fact that crack width may not be controlled by reinforcement, and such cracks may often pass through the element cross-section. Control of this form of cracking is usually done on a 'design out' basis, i.e. by an appropriate material specification and good workmanship, to prevent the cracks occurring or becoming significant.

Possibly, the best known example of extrinsic cracking is that due to direct loading and this is controlled by serviceability requirements in codes of practice, with respect to reinforcement diameters and spacings. Another form of extrinsic cracking in the presence of restraints may arise due to temperature, shrinkage and creep, or to differential movements. 'Designing out' in material terms is not an option here; steps have to be taken to either eliminate the restraints (by providing properly detailed joints) or to accept their presence and design for their effects (usually by providing reinforcement to control any cracking).

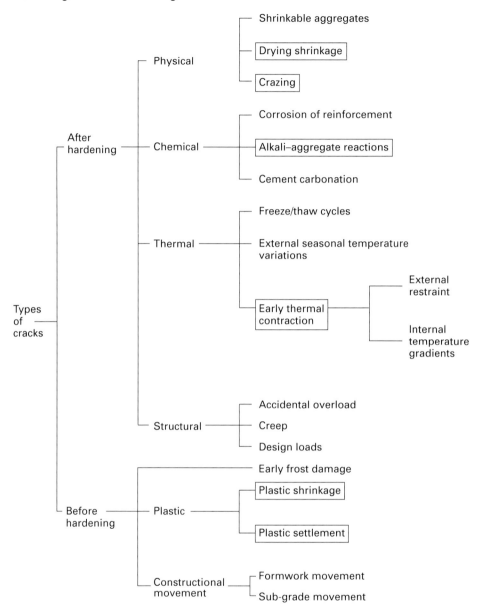

Fig. 1.11 Possible causes of cracking in concrete

It is important to get all of this right. The presence of wide cracks (those with widths well in excess of the normal limits of 0.3–0.4 mm) can increase, by several orders of magnitude, the processes of carbonation and chloride ingress.

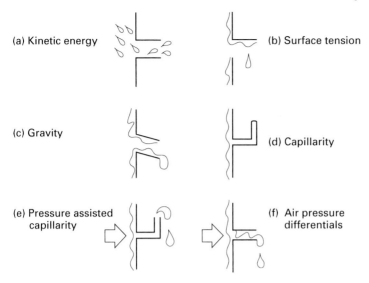

Fig. 1.12 The processes by which rainwater leaks through joints[1.12]

These remarks also obtain to the opening up of badly detailed joints, especially in façades.

Dealing with Water

The presence of water is a prerequisite for many types of deterioration in structural concrete, and its exclusion prevents many problems from occurring. The treatment of water is as critical to durability as that of wind is to lateral stability. Wind and rain, together, can drive moisture through any planes of weakness, particularly in façades – at joints, windows etc., or through any openings or cracks which may have occurred due to different kinds of movement. The mechanisms involved have been identified,[1.12] as illustrated in Fig. 1.12; the simple recognition of these is an essential first step to developing good details at joints.

More generally, both architectural and engineering detailing will influence the interaction between microclimate and the surface of the façade. Minimizing the uptake of water in the material is essential as well as preventing its ingress through joints. A very simple example is given in Fig. 1.13. Run-off, driving rain,

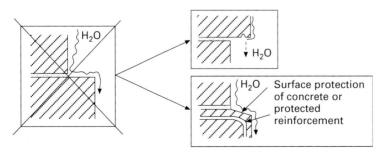

Fig. 1.13 Minimizing uptake of water on a projection

splashing, standing water, or even water vapour are the issues. Depending on the architect's concept of how the building should weather, the options available in very general terms are as follows.

- Protect the building as much as possible.
- Detail, to throw water clear of sensitive areas.
- Collect the water, and dispose of it through a drainage system.
- Detail the façade, to guide the water over the surfaces in a controlled manner.
- Identify sensitive areas where water may penetrate, and develop a redundant protection system – *especially at joints*.

References

1.1   Blanchard B F 1978 *Design and Manage to Life Cycle Cost* M/A Press, Portland, Oregon

1.2   Bennett D F H 1989 *Advances in Concrete Construction Technology* British Cement Association, Crowthorne, UK, Publication No. 97.309

1.3   Paterson A C 1984 The structural engineer in context (Presidential address to the Institution of Structural Engineers, 4 October 1984). *The Structural Engineer* **62A** (11): 335–342

1.4   Somerville G (ed) 1990 *The Design Life of Structures* Proceedings of the Henderson Colloquium, Pembroke College, Cambridge, UK, Blackie, Glasgow

1.5   British Standards Institution 1992 *Guide to Durability of Buildings and Building Elements, Products and Components* BS 7543:1992, BSI, London

1.6   White K H 1991 Building performance and cost-in-use. *The Structural Engineer* **69** (7): 148–151

1.7   Somerville G 1986 The design life of concrete structures. *The Structural Engineer* **64A**(2): 60–71

1.8   British Standards Institution 1985 *Structural Use of Concrete: Part 1: Code of Practice for Design and Construction* BS 8110, BSI, London

1.9   Comité Euro-International du Béton (CEB) 1983 Durability of concrete structures, CEB-RILEM International Workshop, Copenhagen (see also Workshop Report. *Bulletin d'Information* No 152, April 1984)

1.10   Comité Euro-International du Béton (CEB) 1985 Draft CEB guide to durable concrete structures, contribution to the 24th Plenary Session of CEB, Rotterdam, June (see also *Bulletin d'Information* No 155, May)

1.11   Comité Euro-International du Béton (CEB) 1989 *Durable Concrete Structures* CEB Design Guide, 2nd edn (see also *Bulletin d'Information* No 182, June)

1.12   Anderson J M, Gill J R 1988 *Rainscreen Cladding – a Guide to Design Principles and Practice* CIRIA, published by Butterworth, London

2 Structural Modelling of Concrete Buildings

R. F. Warner

Structural modelling is an essential step in the analysis and design of concrete buildings. Mistakes and inadequacies in structural modelling can lead to serious design defects, and to structural inadequacy and failure. This chapter discusses the idealizations and simplifications which are commonly made in modelling concrete buildings.

Highly simplified structural models can be used to obtain very simple approximate estimates of important design quantities such as the bending moments and shears in critical components of a building. Approximate methods are very useful in the preliminary stages of design. They also allow independent, simple order-of-magnitude checks to be made on computer calculations and on the work of other designers.

2.1 Structural Modelling

Structural modelling is the process of creating a simplified, abstract representation of a real building structure. The resulting structural model provides the means whereby the behaviour of the real structure can be investigated analytically, using the principles of structural mechanics.

Real concrete buildings are extremely complex when observed closely, and in practice it is not possible, or even desirable, to work with highly accurate models which mirror all the complexities of the real building. Idealization and simplification are therefore essential elements in the modelling process. To be useful, a structural model must nevertheless capture the essentials of structural behaviour, and in particular indicate the way the structure channels the applied loads into the foundations.

Structural modelling is an essential step in both analysis and design. Errors and inadequacies in the structural model can result in serious design defects, and can lead to structural inadequacy and failure. The potential for serious error in the modelling process increases directly with the complexity of the building structure. Particular care therefore needs to be exercised in modelling large concrete buildings.

The Role of Structural Modelling in Analysis
Structural analysis is used to predict the behaviour of a real structure under prescribed conditions of load and environment. The first step in the analysis is to identify the underlying structural system and its components, and hence create an idealized, abstract model which can be used as the object of the analysis.

The form of the structural analysis depends on the shape, slenderness and complexity of the model components (line, planar or massive elements), and on the assumed material behaviour (linear or non-linear). The analysis is undertaken using the principles of mechanics. Today, with the ready availability of structural software packages and cheap computing facilities, the detailed computations are undertaken by computer.

A final and important step in the process of analysis is interpretive: the numerical results obtained from the analysis of the model are used to estimate the behaviour of the real structure. Due allowance must be made for the inevitable differences between the model and the real building, and possibly also for differences between the assumed loads and environmental conditions and those which actually occur.

The process of structural analysis as shown in Fig. 2.1 thus involves:

- the creation of a simplified, idealized structural model;
- analytical computations for the model;
- interpretive transfer of the results of the analysis of the model back to the real structure.

At the time when the analysis is undertaken, the building might in fact exist only on paper and in the mind of its designer. Alternatively the analysis might be part of an investigation of an existing structure. In either case, the structural analysis of a building involves the essential step of model creation by means of idealization and simplification.

Given the ready availability of computer software packages, one might expect that it would be a simple matter to choose software to suit the particular model

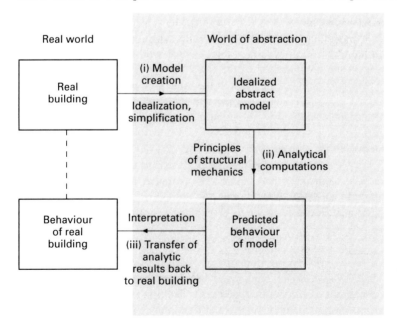

Fig. 2.1 Structural analysis and the role of modelling

which has been created for a real structure. In fact, the reverse is usually the case, and a force-fit of the model to the software often has to be made. For example, slab–column buildings are frequently analysed using simple linear frame analysis packages, because of their low cost and ready availability. To employ such packages, the designer must create a highly idealized stick model consisting of beams and columns, with the two-dimensional slabs replaced by 'equivalent' beams in the principal directions of the building.

Although structural models are usually created in an abstract, mathematical format suitable for computer calculation, there are other possibilities. For example, a scaled-down simplified physical version of the real structure can be constructed in the laboratory, and investigated experimentally. Physical modelling was popular in pre-computer days, and still finds occasional use in the investigation of complex building structures. The important point here is that, irrespective of the form of the model, the overall process of analysis is unchanged. Firstly, a simplified model of the real structure is created, and is then studied quantitatively. The quantitative results of the model study are used to estimate the behaviour of the real structure, with appropriate interpretations being made to allow for the differences between the model and the prototype.

The Role of Structural Modelling in Design

Structural modelling plays an essential role in both the preliminary concept phase of design and in the iterative detailed phase.[2.1] In the preliminary phase the designer cannot afford to work with the structure in all its complexity, and therefore creates a highly simplified conceptual model of the basic structural system. In the detailed stage, relatively precise results are required and a more accurate model has to be used as the object of the structural calculations. The central role of modelling in the detailed iterative stage of structural design is illustrated in Fig. 2.2.

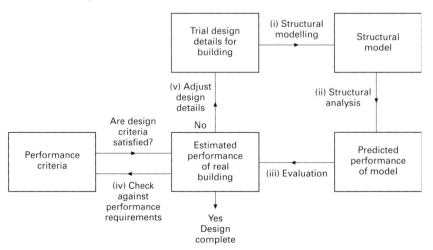

Fig. 2.2 Structural modelling in the detailed design process

The designer thus needs to be able to model at various levels of accuracy or complexity. The success or otherwise of the final design depends in no small part on the adequacy of the structural modelling. Experience, understanding and structural intuition are very desirable attributes for anyone undertaking design in general, and structural modelling in particular.

2.2   The Modelling Process

The creation of a structural model involves a progressive move away from the complex real building towards a simplified idealization. Conceptually, this requires a stripping away of the structurally non-relevant and secondary features of the building in order to expose the components of the underlying load-resisting system, with its internal load paths. Each part of the building needs to be considered in turn, to evaluate its interaction with other adjacent parts and hence its contribution, if any, to structural action.

It is important to consider the connections where the structural components meet, and decide whether or not they are rigid, or whether they deform and allow only limited moment transfer. Joints and connections obviously become a very important consideration if the building contains precast elements.

In concrete buildings a primary structural system can usually be readily recognized, together with the load paths it provides. However, closer inspection can reveal subsidiary structural actions with a variety of additional, but complex, load paths. Whether or not these need to be included in the structural model depends on the level of accuracy required.

2.3   Model Accuracy

In the case of a simple concrete structure such as an isolated portal frame or a continuous beam it is possible to create a very precise structural model. If this model is analysed using an accurate computer program which takes proper account of complex 'second order' effects such as non-linear material behaviour, geometric non-linearities and deformations within the joints, the behaviour of the real structure under a specific set of loads can be predicted with very good accuracy.[2.2]

However, such high accuracy is only achievable with a considerable expenditure of time and computational effort. This may be justified in the case of a research investigation, but it is rarely warranted in practice. In the design of complex building structures, computational accuracy is severely reduced by the practical need to disregard complex 'secondary' effects such as tension stiffening, joint flexibility, torsional and shear deformations, and the many secondary load paths brought into play by the presence of cladding, interior walls, functional attachments and other 'non-structural' elements. The accuracy of the design calculations is further diminished by lack of precise input information on the future building, concerning the as-constructed details of geometry, steel location in the sections, material properties, and the actual loads and environmental conditions.

Generally speaking, the structural models which are used in design practice are very approximate. Despite the appearance of precision given by computer computations, the uncertainties and variabilities inherent in the input data and the idealizations and simplifications which are introduced in the structural modelling ensure that predictions of structural behaviour and strength are far from being precise. They should always be treated with care, and healthy scepticism.

2.4   Simplification and Idealization in Structural Modelling

At the commencement of the modelling process it is important to consider the characteristics of the available computational software packages. The choice of software should be linked to the modelling process. The accuracy and reliability of the design data also need to be considered carefully if sensible decisions are to be made in the modelling process. The aim is always to balance desired simplicity against required accuracy. Idealizations and simplifications are introduced into the modelling process to achieve the desired balance.

Unfortunately, many of the modelling approximations which are sanctioned by design codes tend to be accepted uncritically and treated as accurate representations of real behaviour. The suitability of any approximation needs to be considered carefully in the specific context of each application. It is particularly important to recognize and identify all approximations and simplifications when, in the final step of the analytical process, the numerical results derived from the model are used to estimate the behaviour of the real structure.

The simplifications and approximations commonly used in modelling concrete buildings are discussed briefly in the following paragraphs. The emphasis here is on idealizations regarding structural action and load paths, rather than on material behaviour and loads and environment, which are mentioned only very briefly.

2.4.1   Simplifying and Approximating Structural Action

Neglecting Secondary Load Paths

Non-structural components such as exterior cladding elements and internal walls and partitions usually add considerably to the stiffness and strength of the structural system by providing secondary load paths. They can be extremely complex and difficult to model. For example, a cladding element or wall may only begin to act structurally after some initial movement closes the gaps between the element and adjacent columns and beams. Such behaviour is highly non-linear, and involves frictional effects which require quite sophisticated modelling. If only for reasons of convenience and simplicity, such secondary effects are usually ignored.

While it is conservative to ignore secondary contributions to strength and stiffness, it is important to recognize that performance problems such as unsightly cracking may arise in non-structural elements if allowance is not made for the deformations imposed on them when in fact they act structurally.

Reducing the Dimensionality of the Structural System

Even in modelling for the detailed stage of design, the real three-dimensional building structure is often replaced by a highly idealized model consisting of a set of parallel two-dimensional frames, each independently carrying a proportion of the horizontal load in the (say, north–south) direction of the frames. A second transverse but independent set of plane frames is assumed to carry the loads in the other (east–west) direction (Fig. 2.3). Such simplifications are necessary if the computations are to be undertaken using a two-dimensional frame analysis program. This idealization is reasonable if the building is regular and rectangular, but questionable if there are irregularities, say in the column grid. Various effects have to be ignored, such as torsion in the overall frame and in individual members, and biaxial bending in columns. Corrective design procedures therefore have to be employed to make up for the deficiencies of the analysis, and it is an important task of the designer to recognize such deficiencies and make allowance for them as appropriate.

A reduction of the three-dimensional system to two sets of plane frames is also made in dealing with vertical load effects. In this case, slab action is again replaced by simple beam bending in two directions.

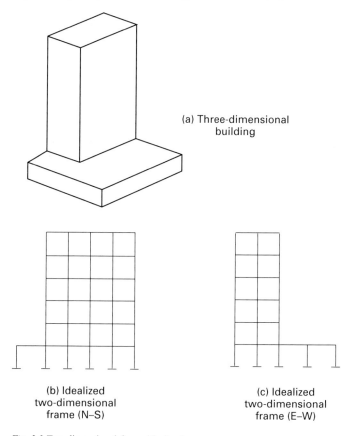

(a) Three-dimensional building

(b) Idealized two-dimensional frame (N–S)

(c) Idealized two-dimensional frame (E–W)

Fig. 2.3 Two-dimensional frame idealization

Despite the obvious approximations involved in two-dimensional models, they have been used successfully for the design of countless concrete buildings. The enormous saving in computational effort will ensure their continued popularity.

Reducing the Degree of Indeterminacy (Using Sub-assemblages for Vertical Loads)

In the analysis of an idealized two-dimensional frame subjected to vertical load, further simplification can be achieved, without great additional loss of accuracy, by reducing the degree of indeterminacy of the model. For example, single-floor sub-assemblages can be analysed in lieu of the entire frame. Such a sub-assemblage consists of one floor and the columns immediately above and below, fixed at their remote ends (Fig. 2.4). The sub-assemblage approach works because bending effects in continuous beams and frames attenuate rapidly with distance from the point of application of a vertical load. That is to say, the moment and shear force at any particular cross-section depend largely on the loads applied in the immediate vicinity. Various design codes and standards even allow sub-assemblages to be used for detailed design calculations.

When vertical design loads are applied throughout a frame, the column forces increase progressively down the building and while it is inappropriate to use sub-assemblages to determine the design axial forces, the column moments produced by the vertical loading can be estimated from sub-assemblage analyses.

Reducing the Degree of Indeterminacy (Locating Points of Zero Moment)

The column moments due to horizontal loads increase in magnitude down the building, because of the cantilever-like action of the structure, and a sub-assemblage analysis cannot therefore be used to treat horizontal load. However, the process of reducing the degree of indeterminacy can be taken a significant stage further by assuming locations for the points of zero moment (contraflexure points). For example in Fig. 2.7 (Section 2.5.2) contraflexure points are placed in the columns at mid-height and in beams at mid-span throughout the frame to treat the case of horizontal loading. This approach provided the main analytical basis for building design in pre-computer times. It is also the basis for the approximate

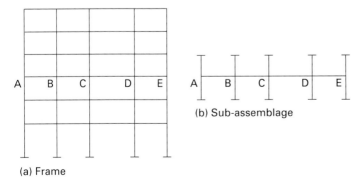

(b) Sub-assemblage

(a) Frame

Fig. 2.4 Frame and sub-assemblage (vertical loading)

methods of frame analysis for horizontal load, which are discussed briefly in Section 2.5. Approximate analysis for the effects of vertical load can also be undertaken by locating the points of contraflexure in a sub-assemblage model such as the one in Fig. 2.4(b).

Neglecting Actions in Elements (Choosing Stiffness Values of Zero and Infinity)

In designing an individual element in a building, the designer may sometimes choose to ignore the effect of a particular action, even though it is known to be present. An example is neglect of torsion in a spandrel beam at the edge of a floor, because it is considered to be small in magnitude, or because experience shows that serious design defects will not result.

The desirable procedure in such instances is to set the torsional stiffness of the beam to zero in the analysis so that alternative load paths are called into play to compensate for zero torsion by increased moments elsewhere, which thus maintain an equilibrium configuration as the basis of the design. If such a modified analysis is not possible because of limitations in the computing software, then a simple estimate of the redistributions can and should be made.

Simplifications can also result from setting stiffness values to infinity. For example, it is quite reasonable to assume rigid diaphragm action in slab floors in order to reduce the degrees of freedom of movement to be considered for the columns in a building. Infinite stiffness values can also be used to separate a local region from the rest of the structural system, in order to create a local sub-assemblage (Fig. 2.5).

Idealizing Joint Behaviour

In most frame analysis packages it is assumed that beam–column joints undergo zero internal deformation and can accept the full end moment from each entering member. Experimental studies show that joint deformations can be very significant and can reduce overall frame stiffness and produce deflections far larger than predicted. The load capacity of the frame may also be much reduced by

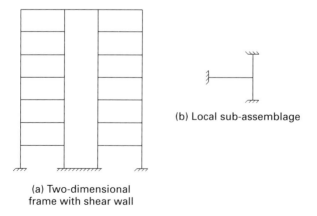

(b) Local sub-assemblage

(a) Two-dimensional frame with shear wall

Fig. 2.5 Use of infinite stiffness to create local sub-assemblage

limited moment capacity within the joints.[2.2] Limited moment capacity in a joint leads to a drastic redistribution of moments in adjacent components. Joint inefficiencies and joint deformations are most serious when there are few entering members, such as in a portal frame. The implications need to be considered in the final interpretive stage of analysis. Clearly in the case of precast construction, the joint assumptions have to be considered carefully at the model creation stage.

Simplifying Load Paths in Local Disturbance Regions
Complex patterns of internal forces develop in local 'disturbance' regions, such as near an applied concentrated force or at a sudden change in geometry in a member. The modelling of local conditions becomes necessary when the design details for the region are determined. A simplified system of internal tensile ties and compressive struts can be used to model the flow of forces through the disturbance region, while equilibrating any externally applied forces in the region. Design then consists of providing tensile reinforcement to carry the tie forces and checking that the concrete can carry the assumed compressive strut forces. This strut-and-tie approach is dealt with in detail in Chapter 5 of this book.

2.4.2   Idealizing Material Behaviour

The design of most concrete buildings is presently based on linear elastic analysis, which is clearly a highly idealized representation of the behaviour of concrete structures. Linear elastic analysis is used not only to investigate working load behaviour, but also to estimate the distribution of the stress resultants (moments, axial forces, shears etc.) under design ultimate load conditions.

Ad hoc corrective procedures must therefore be employed to allow for important non-linear effects such as cracking, tension stiffening at working load, and lateral column deflections under design ultimate load. Adjustments also may need to be made to allow for inelastic redistribution of the internal moments under overload conditions. Other time-dependent effects in the concrete such as creep and shrinkage are often ignored in analysis, and again corrections may have to be made, for example to estimate additional long-term deflections and column shortening.

2.4.3   Geometric Idealizations

Some idealizations are used so often in structural modelling that they tend to be forgotten by some designers. An example is the assumption that the cross-sectional dimensions of the beams and columns in a frame are negligible in comparison with other geometric quantities such as storey height and column spacing. While this assumption is usually acceptable, difficulties arise if the frame contains a transfer beam with a height comparable with the storey height, or a shear wall with an overall dimension approaching the spacing between columns. Special techniques can be introduced to adjust for finite member size if simple frame analysis is to be retained. For example, rigid arms can be introduced between the axis of the large member and adjacent 'normal' members to allow for member size and the increased frame rigidity. Such methods are discussed in Chapter 3 of this book.

First-order frame analysis procedures assume that equilibrium requirements can be formulated for the structure in its original, undeformed position. Concrete buildings and components tend to be stockier than their steel counterparts, and second-order geometric effects are correspondingly less significant. However, even when local deformations in a column are small and negligible within a storey, the overall lateral deflections in a multistorey building frame need to be considered in the formulation of equilibrium requirements. Most design codes and standards allow for a simplified treatment of the P–Δ effect in individual compressive members, although a non-linear analysis of geometric effects leads to a conceptually more satisfactory design treatment.[2.3]

2.4.4   Idealizing Loads and Environmental Effects

Precise information on the magnitude and spatial and temporal distribution of loads is rarely available at the time the design is carried out. The idealized design loads are therefore treated as being either uniformly distributed or concentrated, and as acting either continuously or instantaneously in time. The design values specified in relevant loading codes for such idealized loads necessarily impose limits on the accuracy with which future structural behaviour can be predicted.

Similar limitations apply to short-term and long-term seasonal variations in environmental parameters such as temperature and humidity, and to foundation behaviour which can usually be predicted only very approximately, even after careful site investigation.

2.5   Approximate Analysis

After a structural model has been created using the appropriate simplifications and idealizations, further approximations can be introduced which reduce drastically the complexity of the model, and hence the computational effort. The resulting 'approximate' methods of analysis allow extremely simple but rough estimates to be made of key design parameters such as the stress resultants in critical regions of a building. Although there may be a significant loss of accuracy because of the simplifications introduced, experience shows that these methods can be valuable, provided they are used with understanding.

The approximate methods have been in existence for many years.[2.4] Prior to the computer they provided the analytical basis for the successful design of many complex concrete buildings. Even with the prevalent use of computer analysis today, the approximate methods can still play a useful role in design. In the preliminary concept stage, alternative structural systems need to be considered and compared, and it is important to be able to undertake simple order-of-magnitude calculations to estimate critical design parameters, such as maximum moments in critical regions, in order to obtain trial member sizes. The approximate methods are ideal for this purpose.

Although graphic input and output can improve computer friendliness enormously, the designer always needs to be able to make an independent, order-of-magnitude check on critical output quantities. The approximate methods of analysis provide this facility. They can also be used to undertake a rough check on calculations of other designers.

An extensive literature exists on approximate methods of analysis.[2.4–2.7] The present discussion is therefore brief.

2.5.1   Approximate Analysis for Vertical Loads

An approximate analysis for a continuous beam or floor in a building frame under vertical loading can be achieved by restricting attention to a sub-assemblage in the immediate region and guessing the locations of the points of contraflexure in the columns and beams (Fig. 2.6). By iteratively sketching both the deflected shape and the moment diagram for each beam, and ensuring that the points of zero moment and zero curvature coincide in the two sketches (Figs 2.6(b) and 2.6(c)), a good estimate of the moment diagram can usually be achieved.

The fixing moments for the beams depend of course on the moments in the columns immediately above and below the floor. Under vertical load the columns in a frame are typically subjected to double bending with the point of zero

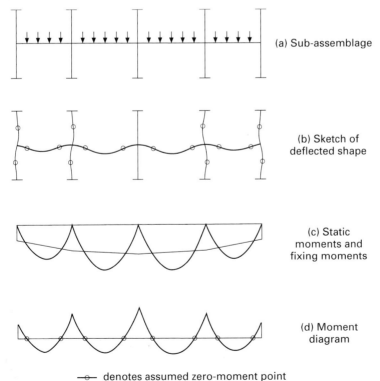

(a) Sub-assemblage

(b) Sketch of deflected shape

(c) Static moments and fixing moments

(d) Moment diagram

—⊙— denotes assumed zero-moment point

Fig. 2.6 Moment estimates in subassemblage (vertical loading)

moment not too far from mid-height. If the frame is regular, with equal or near equal spans, the moments in the interior columns are very small and can be ignored. In this case the maximum positive and negative moments in a beam in an interior span are obtained simply by isolating the beam element and treating it as built in at each end.

In the end span the fixing moments depend on the stiffnesses of the exterior columns. A reduced sub-assemblage similar to the one shown in Fig. 2.5(b) can be used, and estimates can be made of the location of the points of contraflexure. The moments in beams and columns adjacent to a core or shear-wall element can also be obtained in a similar way.

If a more accurate procedure is desired for estimating moments by means of inflection points, use may be made of various proposals which have been recently published.[2.5,2.6] A procedure was recently suggested by Epstein[2.7] for accurately locating inflection points which is relatively simple to apply. Check calculations suggest that Epstein's method is accurate and reliable for a wide range of frames.

2.5.2   Approximate Analysis for Horizontal Loads

Two approximate methods which have been used for many years for the analysis of frames subjected to horizontal load are the *portal method* and the *cantilever method*. In each method the degree of statical indeterminacy is greatly reduced by assuming points of zero moment at the mid-points of beams and columns (Fig. 2.7).

In the portal method the additional assumption is made that the shears in all interior columns at any level are equal, and twice as large as in the two exterior columns. This simplification works well in the upper regions, but usually needs adjustment for the lowermost floors.

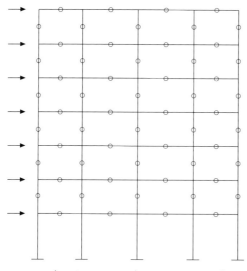

—o— denotes assumed zero-moment point

Fig. 2.7 Moment estimates in frame (horizontal loading)

The cantilever method also assumes contraflexure points; however, the axial forces in the columns are calculated by treating the entire frame as a simple cantilever with its cross-section made up of the individual column areas. In a frame with equal column sizes the axial forces vary with distance from the centre of the frame, with the maximum tensile and compressive forces occurring in the outermost columns.

The cantilever method becomes progressively more suitable as the frame becomes higher. As both methods ignore the relative stiffnesses of the beams and columns, results can be obtained which are significantly different from an elastic analysis. However, the relevance of a linear elastic solution as the basis for comparison is itself open to question, since reinforced concrete is neither linear nor elastic.

2.5.3   Other Structural Systems

The approximate methods apply to simple structures such as frames and continuous beams, and are not directly applicable to more complex concrete buildings. Nevertheless, simplifications and approximations can be introduced at the initial modelling stage in order to create a highly idealized model for simplified calculations. The assumptions clearly must take account of the flow of forces through the structure.

Idealizations can often be made to create simplified sub-assemblage models for local regions of the structure. Points of contraflexure, chosen with the assistance of a sketch of deflected shape, can then be introduced as a means of estimating actions in critical regions.

2.6   Structural Modelling and the Theory of Plasticity

Idealization and simplification eventually lead to structural models which are gross distortions of the original building structure. The two-dimensional model (Fig. 2.3) is a case in point. Designers obviously need to be aware of the effects of the idealizations in their structural models, and in particular of whether or not the resulting calculations are conservative.

The overload behaviour of a concrete structure can be extremely complex and strictly speaking is neither linearly-elastic nor elastic–plastic.[2.2] Modelling errors can therefore only really be investigated by comparative means. That is to say, the errors introduced in the creation of a specific structural model can only be estimated by repeating the analysis using a more accurate model.[2.8] Studies of the accuracy of the simplified models commonly used in design have become possible with the development of sophisticated non-linear computational software programs.[2.2,2.8,2.9]

Nevertheless, reinforced concrete can be surprisingly ductile if care is taken in choosing the amount, location and distribution of the tensile reinforcement. The resulting behaviour at high overload can then approach the elastic–plastic ideal if the system is stable, robust (not prone to progressive failure) and ductile. In such circumstances it is reasonable to apply the principles of plasticity theory to the overload behaviour of concrete structures.

Of particular interest in reinforced concrete design is the lower bound theorem of plasticity,[2.10,2.11] which can be expressed non-rigorously as follows.

If a set of internal stress resultants (moments, shears etc.) can be found such that:

- it equilibrates the external applied loads and
- the stress resultants in each local region are not large enough to cause failure,

then the structural system can safely carry the applied loads.

In the context of design, this means that a structural system will safely carry a specified ultimate load if it is designed for any distribution of internal stress resultants which is in equilibrium with that load, and if all sections and regions of the system are provided with sufficient material to carry the assumed stress resultants. The key word in this statement is *any*. The designer does *not* have to determine the *actual load paths* in the structure under ultimate load conditions in order to achieve a safe design. The design may be based on any convenient load path which equilibrates the applied load.

This supports the view that even gross idealizations of the real structural system, such as the one shown in Fig. 2.2, will lead to a safe design, provided the designer aims at achieving ductility and makes sure that the overall requirements of equilibrium are satisfied. Simplifications such as reducing the degree of indeterminacy by guessing the locations for points of contraflexure should also result in a safe design provided the design moments are in equilibrium with the applied loads.

It must be emphasized that the plasticity considerations introduced here deal with equilibrium conditions and structural behaviour 'in the large'. There can be no guarantee that local effects, such as bond breakdown between concrete and reinforcement, will not lead to structural failure. In localized regions, structural integrity often depends on the ability of concrete to carry tensile stress. Local effects have to be dealt with through good detailing. The strut-and-tie concepts can play a useful role here.

Clearly, the considerations of plasticity are not relevant to serviceability design. Indeed, if the equilibrium configuration assumed as the basis for strength design is markedly different from the equilibrium configuration actually developed under low-load conditions, then a significant amount of redistribution of the internal stress resultants may occur as the load increases. Redistribution, or modification of the load paths, is usually accompanied by a significant amount of cracking and this can lead to serviceability problems, including excessive cracking. Such problems can be largely avoided by the choice of equilibrium configurations which are not too different from those applying at low loads.

2.7   Concluding Remarks

Structural modelling is a crucial step in both the analysis and design of concrete buildings. Various methods of approximation and idealization have to be used in the modelling process to achieve the desired balance between required accuracy and desired simplicity.

Structural modelling requires a good understanding of structural principles as well as a good structural intuition. Unfortunately, the absence of these qualities in a design team will not be overcome purely by the employment of sophisticated computer software. Indeed, the likelihood of design error tends to increase with increasing sophistication of computer software, especially when a mature understanding of structural principles is lacking.

On the other hand, highly simplified models can give an insight into structural behaviour. When used with understanding they can provide the basis for the simple calculations which are required in the early stages of structural design. They also provide a useful means for making order-of-magnitude checks on computer computations, as well as on the work of other structural engineers.

References

2.1   Warner R F, Rangan B V, Hall A S 1989 *Reinforced Concrete* 3rd edn, Longman Cheshire, Melbourne

2.2   Kenyon J M, Warner R F 1992 Refined analysis of non-linear behaviour of concrete structures. *Civil Engineering Transactions, Institution of Engineers, Australia* **CE 35**(3): 213–220

2.3   Bridge R Q 1994 Introduction to methods of analysis in AS 4100-1990. *Steel Construction Journal, Australia Institute of Steel Construction* **28**(3): 1–9

2.4   Benjamin J R 1959 *Statically Indeterminate Structures* McGraw-Hill, New York

2.5   Behr R A, Goodspeed C H, Henry R M 1989 Potential errors in approximate methods of structural analysis. *ASCE Journal of Structural Engineering* **115**(4): 1002–1005

2.6   Behr R A, Grotten E J, Diwinal C A 1990 Revised method of structural analysis. *ASCE Journal of Structural Engineering* **116**(11): 3242–8

2.7   Epstein H I 1988 Approximate location of inflection points. *ASCE Journal of Structural Engineering* **114**(6): 1403–1413

2.8   Wong K W, Warner R F 1988 Analysis and design of slender concrete frames by computer. *Proceedings, 2nd International Conference on Computer Applications in Concrete* Singapore Concrete Institute

2.9   Scordelis A 1984 Computer models for non-linear analysis of reinforced and pre-stressed concrete structures. *PCI Journal* (Nov–Dec): 116–135

2.10   Drucker D C 1961 On structural concrete and the theorems of limit analysis. *Memoires IABSE, Zürich* **21**: 49–59

2.11   Chen W-F, Drucker D C 1969 Bearing capacity of concrete blocks or rock. *ASCE, Engineering Mechanics Division* **95**(4): 955–978

3   Computer-Aided Analysis and Design

M. C. Griffith

This chapter presents an overview of the use of computers and various computerized analysis techniques. The components of a generic analysis/design package are first outlined. Specific examples of some of the many commercially available packages are then discussed, as are the general trends in software development. The main aim of this chapter is to give practising engineers who are not already familiar with this topic a basic understanding of the capabilities, and limitations, of commercial software and to introduce them to some of the most widely used methods for computer analysis of concrete buildings. In particular, techniques for modelling beams, columns, slabs, shear walls and connections between these components of concrete buildings are presented.

3.1   Overview of Software Packages and Their Uses

Engineering software for structural analysis has evolved from being something of an oddity a decade ago to being a standard tool used in almost every design office. This is due largely to the fact that software and hardware costs have continued to decrease. One reason for the decrease in software costs is the growing number of commercial packages available which has resulted in a very competitive buyer's market. The competition has resulted both in falling prices and ever-increasing software capability. Furthermore, technological advances in computer hardware have made it feasible to conduct quite complicated structural analyses on small, inexpensive computers. This has enabled many small firms to benefit from the advantages of computerized analysis.

For example, it is now possible (in 1995) to purchase finite-element analysis software with 3D, non-linear, static and dynamic capabilities to run on desk-top personal computers for under US$2000. Not so long ago this type of software was priced at tens of thousands of dollars and was only available to run on large, mostly mainframe, computers. Now there is finite-element software which can analyse just about any structure for every type of loading imaginable. It is also possible to run most commercial packages on any type of computer, from lap-top personal computers (PCs) to large mainframe (DEC, IBM, etc.) multi-user computers. Undoubtedly, much of the analysis software currently available is still mostly applicable to researchers. The characteristic of research software is that it tends to have enormous computing power and modelling capabilities but is not generally 'user-friendly', i.e. easy to use. In contrast, the most popular commercial software is, in general, easy to use.

The basic structure of a generic analysis package is shown in Fig. 3.1. Of course, every commercial package will not have all of the analysis capabilities indicated in Fig. 3.1. However, all packages have three common parts: (1) a 'preprocessing' module, (2) an analysis module and (3) a 'post-processing' module. Historically, the analysis module has been the primary focus of attention of software developers and consists of load definition modules and modules to solve for displacements, element forces and stresses, and support reactions. The type of loading dictates the type of analysis which is required.

Since the mid 1980s, much of the commercial software development effort has shifted from the analysis modules and been put into the improvement of 'pre-processors' and 'post-processors' (Fig. 3.1). 'Pre-processors' are modules (or computer programs) which allow engineers to easily define a numerical model for a given building and convert this information into a format which can be operated on by the structural analysis module. Similarly, 'post-processors' are programs which take the results of a structural analysis, joint displacements or member forces for example, and manipulate them in some way to make the results more easily interpreted by the engineer. Examples of post-processors include programs which plot the deflected shape of the structure or plot shear force and bending moment diagrams. In fact, the last 10 years has seen a rapid expansion in post-processing programs which take the results of a structural analysis and check each element against the requirements of relevant engineering codes for strength, serviceability, or other design requirements, and in some instances, propose alterations where members do not satisfy specific design requirements. In addition, there is also a wide variety of computer-aided drafting packages (AutoCad, for example) which can be used in varying degrees as the starting point for an engineering analysis. Many of today's analysis programs include pre-processors and post-processors and also come with 'translators' which will convert, for example, any AutoCAD drawing file which has been saved in the .DXF format into a format which can be used as the starting point for the model definition of the mathematical model of the structure to be analysed.

It should also be pointed out that design modules, or at least code checking modules, are slowly being incorporated into many of the more traditional analysis packages. A flow chart of a generic analysis/design package which has this ability is shown in Fig. 3.2. The thought that this process might one day become completely automated, thereby eliminating the need for structural engineers, is worrying. However, experience indicates that even the most sophisticated design/analysis packages still require a skilled structural engineer to 'drive' the analysis. This matter is also discussed in Chapter 2 of this book.

A list of some of the PC-based finite-element/structural analysis programs which are currently available is provided in Table 3.1. Prices for packages similar

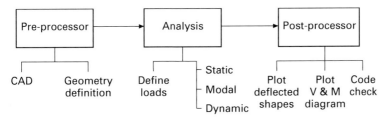

Fig. 3.1 Flow chart for a generic structural analysis software package

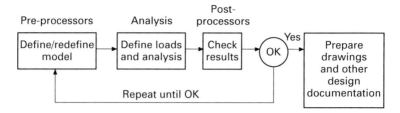

Fig. 3.2 Flow chart for a generic design/analysis software package

to those listed in Table 3.1 vary between $1000 and $10 000. Inspection of the programs in Table 3.1 indicates that, for the most part, there are few significant technical differences between the software packages. Probably the most important feature of software packages is ease of use. This characteristic is strongly related to the level of integration between the analysis modules, the pre- and post-processing modules, and the design modules. In general, if each of the modules is developed by the same company, then the integration is typically of a high order and the interface between each module is transparent. On the other hand, inter-faces between modules developed by different software companies tend to be restricted and cumbersome, leading to a lower degree of user-friendliness.

Until recently, most designers conducted a detailed structural analysis only as a final check on their design. However, there is a clear trend in the engineering software industry toward the development of fully integrated design and analysis software packages. Indeed, some software companies already claim to offer such software. Hence, in the near future, well-integrated suites of engineering software will be widely available where the structural model can be defined using gra-

Table 3.1 Summary of structural analysis software features

Software package	Features				
	Non-linear, inelastic capabilities	Finite element analysis capability	Design	Integrated in-terfaces with design and CAD modules	Size restrictions
IMAGES-3D	Some	Yes	Code check	Some	Some
NISA-II	Yes	Yes	No	Some	Some
SAP-90	Yes	No	Code check	Yes	Some
COSMOS/M	Yes	Yes	Code check	Yes	Some
STARDYNE	Yes	Yes	Yes	Yes	None
SCADA	Yes	Yes	Yes	Yes	Some
ANSYS	Yes	Yes	Code check	Yes	None
STRUCDES	No	No	Yes	Yes	Some
ACES	No	No	No	Some	Some
SPACEGASS	No	No	Yes	Yes	None
STRAND6	Yes	Yes	No	Yes	None
MICROSTRAN	Yes	No	Yes	Yes	None
NASTRAN	Yes	Yes	Yes	Yes	Some

phical input, possibly computer-aided-drawing (CAD) software, and then designed and analysed iteratively using other software modules. The design will be modified after each iteration to comply with the relevant material, loading, and any other design code requirements which apply. Once the design is finalized, joint details and drawings will be automatically produced by the computer along with all the necessary design documentation and calculations. Eventually, expert system programs may be incorporated into this suite of programs to add on-line advisory capabilities which can rapidly direct designers to an economical and feasible design solution.

The following sections of this chapter will discuss various types of analysis associated with the design of large concrete buildings, and in particular: static and dynamic analysis; linear and non-linear analysis; and two- and three-dimensional analysis. Methods for modelling frame and shear-wall components and their interaction are also covered. The chapter concludes with a summary of some recent developments in the field of computerized structural analysis and the most likely future developments.

3.2   Review of Analysis Range

When considering the general topic of computer-aided analysis of reinforced concrete buildings, the question of what type of analysis is required comes to mind. The answer depends upon many things including the type of structure, the stage of design, type of loading and the information required. For example, the questions of whether to conduct a static or dynamic analysis, approximate or 'exact' analysis, two-dimensional or three-dimensional analysis, linear or non-linear analysis must all be answered before embarking on the analysis and design of a building.

3.2.1   Types of Analysis

Static versus Dynamic

The method of analysis used and the corresponding mathematical model required is dictated by many things, not least of which is the type of loading. Loads can be divided into two broad categories: static or dynamic. Static loads may consist of concentrated forces (or point loads) and distributed loads which are typically due to the structure's self weight plus the design live load. Dynamic loads due to wind and earthquake cause movements in buildings which vary as a function of time. However, designers are mainly interested in the maximum responses of the building to each type of load, such as axial and shear force, bending moment and deflections. Consequently, the effects of wind and earthquake loads on buildings have traditionally been estimated by static analysis techniques using approximate static force representations for wind and earthquake loadings. Even the modal response spectrum method of dynamic analysis which is sometimes used for earthquake design is essentially a method of distributing an equivalent static force throughout the building in proportion to the mode shapes and combined with the results of other modes to give an estimate of the maximum response of the building to a design earthquake.[3.1] This method has proved to be satisfactory, but in certain cases where key structural components are sensitive to cyclic loading, full dynamic analysis is still required. Furthermore, with the increased use of high-strength concrete, structural vibrations and building flexibility are becoming more of an issue requiring analytical checks of the vibrational characteristics of the buildings.[3.2]

Preliminary versus Final Analysis

At the preliminary stages of design, a reasonably accurate estimate is required for the distribution of forces throughout a building (say within 20 per cent or so) without having to perform a time-consuming analysis. For this reason, many approximate methods of analysis have been developed over the years,[3.3] many before powerful and efficient computers were widely available to the practising engineer. Some of these are discussed in Chapter 2 of this book. However, with some insight, an engineer can now use the same computer software packages available for detailed final analysis for preliminary analysis but using a more crude, approximate model of the final design which can be quickly defined and analysed on the computer. For example, a three-storey portal frame structure could be modelled assuming one set of material and cross-section properties for all members of the structure in order to estimate the member actions. The member sizes can then be refined and a second analysis carried out to check the adequacy of the members to safely resist the applied loads.

Two-dimensional versus Three-dimensional

The question of whether to model the building in two dimensions (2D) or three dimensions (3D) may arise when considering the effects of horizontal forces on a building or when the vertical loads are distributed irregularly. This may occur in buildings with irregular floor shapes or in buildings with irregular column spacings. In general, two-dimensional analysis is adequate to capture the essential features of a building's response to both static and dynamic horizontal loads as long as the structure is symmetric in plan and the loading is also symmetric. On the other hand, a full three-dimensional dynamic analysis must be used to account for the torsional response which results in an unsymmetric structure.

Linear versus Non-Linear

Most design work today is still carried out using linear elastic analysis and the wealth of computer software which is now available to do such analyses reflects this fact. Nevertheless, with the rapidly increasing speed and power of computers there is an ever increasing trend among software developers to provide non-linear analysis capabilities by allowing for the definition of non-linear material properties and accounting for second-order effects such as those due to the geometric non-linearities (P–Δ effects) and other large deformation analyses. It is expected that with the widespread adoption of limit state design procedures all analysis will eventually use non-linear methods.[3.4]

3.2.2   Loads

Wind Loads and Aerodynamic Response (Human Perception of Motion)
As mentioned above, wind loads are a form of dynamic load to which tall
buildings are especially prone. There are now generally accepted threshold levels
of motion which humans can perceive.[3.5] The designer must ensure that the
building does not exceed these levels in order to make the building inhabitable.
One way of doing this is to perform a dynamic analysis to estimate the fundamental frequencies of vibration for the building and the corresponding magnitude
of dynamic motion. By adjusting the mass and stiffness, a designer can create a
building which has the desired vibrational characteristics.

Earthquake Forces
Buildings in California, Japan and New Zealand have been designed to resist
earthquake forces for quite some time.[3.6] Today, however, engineers in many
other less seismically active areas such as the United Kingdom and much of
Europe, the central and eastern United States, and Australia are also being asked
to check their designs for earthquake resistance.[3.7] In these parts of the world, the
magnitude of the wind force for a tall building (say > 15–20 storeys) will
probably be greater than the earthquake force. However, earthquakes induce
inertial forces at the centre-of-mass of a building which, if the building is not
symmetric in plan, can lead to significant torsional response. This difference from
wind-load effects makes it an important design consideration.

Vertical Dead and Live Loads
Most of the techniques available for estimating the distribution of forces in a low-
rise building due to vertical loads are equally applicable to the design of tall
buildings and so not much discussion is given in subsequent sections to this topic.
However, with the increasing ease of use provided by modern computer packages, it is expected that the building responses to both horizontal and vertical
loads will be calculated using the same computer model and, where required, their
effects combined using a 'post-analysis' module which is part of the suite of
analysis programs.

3.3   Linear Elastic Two-Dimensional Analysis: Uses and Limitations

The limitations on the use of linear elastic, two-dimensional analysis are related
to the assumptions upon which it is based.[3.8] Some of the most common assumptions are as follows.

1. Concrete behaves linearly and elastically so that building and member
    actions for different load cases can be added algebraically.
2. Floor slabs act as horizontally rigid diaphragms to ensure the horizontal
    displacement of each point at that level of the building are the same.
3. The out-of-plane action of shear walls and slabs can be ignored.
4. $P–\Delta$ effects are neglected (although they do not have to be).

5. Axial and shear deformations can be neglected (although they do not have to be).
6. The effect of cladding and other non-structural components is negligible.

With these assumptions, building structures which are symmetric in plan and which are loaded symmetrically can be analysed using two-dimensional models since each assembly in the vertical plane (frame, shear wall, column etc.) can be expected to deflect identically. Symmetric three-dimensional buildings can be broken down into their two-dimensional lateral force-resisting components and analysed separately or together, in series, as shown in Fig. 3.3 where the frames are linked together with pin-ended axially rigid bars to enforce equal horizontal displacements of the frame. Moment-resisting frames (MRFs) which are used to provide the lateral force resistance of a tall concrete building can be quickly and easily defined and analysed using modern computer packages.

The vertical cantilevered shear wall is another method for resisting horizontal forces. If the height-to-width ratio of a shear wall is sufficiently great, say > 5 (which is the case for most buildings), the shear wall can be modelled as a 'wide column' using an ordinary beam element having the appropriate cross-section properties. However, the horizontal connecting beam elements require rigid-end regions in order to model the effectively rigid width of the shear wall. Otherwise, significant errors will be introduced. Beam elements with rigid-end offsets, or eccentricities, can be used for this purpose (Fig. 3.4).

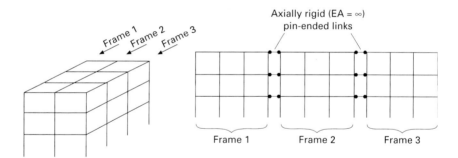

Fig. 3.3 Two-dimensional representation of a frame building

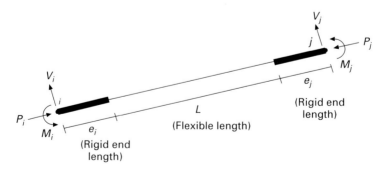

Fig. 3.4 Beam element with rigid end zones

Membrane elements can also be used to model the shear wall, but care must be taken to ensure that the proper rotations are induced into the horizontal beam elements attached to the wall. This problem arises because membrane plate elements have no rotational stiffness at their corners. Hence, a beam element which is connected to the corner of a membrane plate acts as though it is 'pinned', i.e. there is no moment. If the beams are connected to the shear wall in the real building with 'moment connections', then additional elements must be introduced to the computer model. Unfortunately, most finite-element analysis software does not automatically compensate for this.

One way to treat this problem is to introduce a fictitious beam element which is flexurally rigid ($EI = \infty$) to the model.[3.9] The fictitious beam element should be located along the edge of the shear-wall element which is connected to the node where the beam of interest is attached (Fig. 3.5). This allows the rotation at that joint, due to the relative displacements of the nodes at either end of the rigid beam element, to be imparted to the end of the horizontal beam element.

To illustrate this point, a seven-storey shear-wall plus moment-resisting frame structure was analysed using several different models (see Fig. 3.6). The effect of these modelling techniques on the results are listed in Table 3.2. Note that the maximum horizontal roof displacements were significantly larger for the model with membrane plate elements. This is because of the rotational flexibility in this model at connections between beam elements and plate elements (note M_j values). Furthermore, there was a wide range of values for the bending moments in the beams at the point where they were connected to the shear wall. Based on this analysis, it is suggested that for most cases, adequate accuracy can be obtained using a 'wide-column' model and beam elements with rigid end zones.

However, this approach is limited to cases where the shear wall is uniform up the building without many cutouts, or discontinuities. In the event that a shear wall is irregularly shaped or has more than 50 per cent of its effective cross-section missing due to cutouts, a finite-element model using membrane elements for the shear wall may be necessary to accurately capture the building's response as well as to accurately predict the areas of stress concentrations in the shear wall. A better alternative, however, may be to analyse the structure for its overall performance using a model similar to that in Fig. 3.6(c) and to then create a detailed model of the shear wall by itself. This model should then be loaded using the results of the analysis of the overall model. This will enable points of extreme

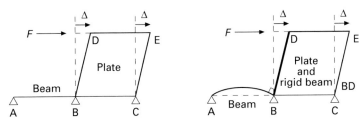

No moment induced in beam without 'rigid' beam

90° angle between beam AB and shear wall preserved by 'rigid' beam BD

Fig. 3.5 Use of rigid beams for moment transfer between plates and beams

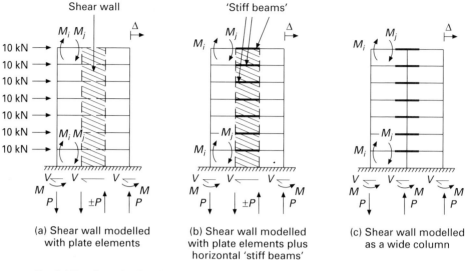

(a) Shear wall modelled
with plate elements

(b) Shear wall modelled
with plate elements plus
horizontal 'stiff beams'

(c) Shear wall modelled
as a wide column

Fig. 3.6 Two-dimensional models for seven-storey frame plus shear horizontal wall structure

Table 3.2 Results of analysis of seven-storey frame plus shear-wall structure

Structural response	Analysis technique		
	Membrane plate without stiff beam at plate edges	Membrane plate with stiff beam at plate edges	Wide column and beams with rigid ends
Lateral roof displacement Δ(mm)	16.0	12.7	12.9
First floor beam end moments M_i/M_j (kN m)	−0.32/0.00	−0.45/−0.60	−0.46/−1.14
Roof beam end moments M_i/M_j (kN m)	−0.54/0.00	−0.65/−1.21	−0.65/−2.13
Base–column reactions $V/P/M$ (kN/kN/kN m)	−0.31/−3.6/0.15	−0.40/−11.6/0.16	−0.42/−11.6/0.17
Base–shear-wall reactions $V/P/M$ (kN/kN/kN m)	−69.5/ ± 175.5/0 ($M_{\text{eff}} = 210.6$)	−69.3/ ± 151.6/0 ($M_{\text{eff}} = 181.9$)	−69.2/0/181.2

stress concentration to be identified in the shear wall without needlessly increasing the computing time for the overall problem.

3.4   Three-Dimensional Analysis

As discussed briefly in Section 3.2, a three-dimensional analysis may be required to adequately capture the response of a structure which is geometrically irregular, either vertically or in plan. For example, earthquake ground motions can induce significant torsional response in buildings where the centre-of-mass and the centre-of-stiffness are separated by a distance e as shown in Fig. 3.7.

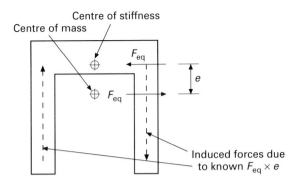

Fig. 3.7 Unsymmetric buildings subject to torsional response (plan view)

The resulting torsion must be resisted by all lateral force-resisting elements at the particular level of interest. The magnitude of torsion can be estimated as the product of the eccentricity e and the horizontal earthquake force, F_{eq}.

While it is possible to resolve unsymmetric loading into symmetric and anti-symmetric components and perform a combination of 2D analyses on symmetric buildings, unsymmetric buildings will almost always require a full 3D analysis. It is also important to remember that when considering earthquake loading, symmetry refers to the mass of the building as well as the lateral force-resisting elements. This is because the inertial forces induced in the building by the ground vibration will effectively be located at the centre-of-gravity of each level of the building. If this position differs from the effective centre-of-stiffness for that level, then the inertial force and the resisting stiffness force form a couple which must be balanced by additional horizontal shear forces in all of the lateral-force-resisting elements at that level. Furthermore, earthquake loading will not only cause a component of torsional response in an unsymmetric structure but its effects may be amplified through modal coupling if the frequencies of the fundamental lateral mode and the fundamental torsional mode are nearly equal. Much has been written about lateral–torsional mode coupling in the response of tall buildings and there are now procedures available for estimating its effect through the use of amplification factors as part of an equivalent static analysis.[3.10] Nevertheless, while wind and earthquake loads are dynamic in nature, only for very special cases such as buildings having post-disaster functions is full dynamic analysis required for design. For most buildings, an equivalent static force method of analysis is adequate.

Another type of building which commonly requires three-dimensional analysis is a building which has non-planar force-resisting elements (Fig. 3.8). It is also not unusual to have members which resist forces in two or more directions such as a column which is part of two separate moment-resisting frames which are oriented at right angles to each other and which have a corner column in common (Fig. 3.9). If the structure is symmetric, then it may be possible to analyse the structure as two separate 2D models and combine the effects in the corner column using the principle of superposition. For the special case of earthquake loading,

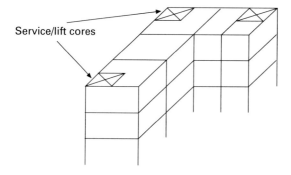

Fig. 3.8 Building with non-planar force resisting systems

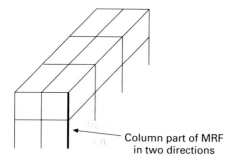

Column part of MRF
in two directions

Fig. 3.9 Moment-resisting frames
with common column members

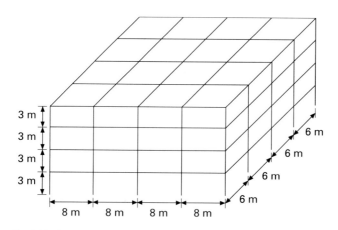

Fig. 3.10 Three-storey moment-resisting frame structure example

Table 3.3 Results of analysis of MRF structure example

No. of storeys	No. of joints	No. of members	No. of DOFs	Matrix bandwidth	Time (s)
2D model					
3	20	27	45	6	25
20	105	180	300	9	39
45	230	405	675	9	69
3D model					
3	100	195	450	22	73
20	525	1300	3000	43	1009
45	1150	2925	6750	43	2809

guidance is normally given in the relevant earthquake codes regarding the combination of results of earthquake loading in orthogonal directions.

A common problem with 3D structural analysis is that the problems are significantly larger to solve than the corresponding 2D problems. Both the model definition is more complex, requiring more time and effort, and the computing time is increased due to the increased number of nodes and corresponding degrees of freedom per node. For example, a simple 3D model of a three-storey MRF having five frames positioned in each orthogonal direction would require considerably more computing effort to solve as a 3D problem over the effort required to analyse it as an equivalent 2D structure. To illustrate this point, three-storey, 20-storey and 45-storey versions of the frame shown in Fig. 3.10 were analysed using a commercial analysis program on a 486 PC. The computing time required to solve for the joint displacements and all member forces and moments are given in the rightmost column of Table 3.3. It can be seen that a 2D model of even a relatively large structure can be analysed quite quickly on a PC (45 storeys analysed in 69 s). On the other hand, the respective 3D models for the same set of buildings took approximately three times (73/25), 25 times (1009/39) and 40 times (2809/69) longer to solve for the corresponding displacements and member forces/moments. Nevertheless, the total time taken to solve the 3D model of the 45-storey building was less than 40 minutes! As the new generation of PCs come into wider use, these figures will decrease greatly, but the relativities are not likely to alter significantly.

A modelling technique which is especially applicable to structural analysis on PCs is the use of master/slave nodes. This feature is now readily available in many computer packages. This method reduces the computational effort for large problems and is based on the realization that there are many examples where certain displacements in a structure are not independent. For example, it is commonly assumed that the floor slab acts as an essentially rigid diaphragm. Thus, if the two horizontal displacements and rotation are known at a single point in the slab, then the horizontal displacements and rotation can be calculated for every other point in the slab (see Fig. 3.11).

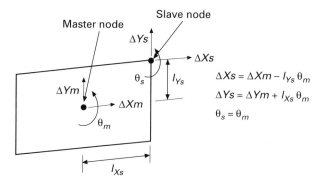

$$\Delta Xs = \Delta Xm - l_{Ys}\,\theta_m$$

$$\Delta Ys = \Delta Ym + l_{Xs}\,\theta_m$$

$$\theta_s = \theta_m$$

Fig. 3.11 Master/slave node relationships for a rigid slab diaphragm

3.5   Non-Linear Analysis

The assumptions commonly associated with linear-elastic analysis of concrete structures were listed previously in Section 3.3. In this section, a brief outline of non-linear analysis is presented. Two types of non-linearities are considered: material (stress–strain) non-linearity and geometric (strain–displacement) non-linearity. In both cases, the non-linearities result in a stiffness matrix which is strain (or displacement) dependent and leads to the use of iterative solution techniques.

It is widely recognized that the stress–strain relationship for concrete is not linear. Furthermore, the effects of creep on concrete structures can be significant and can be accounted for using non-linear analysis techniques. In fact, both of these problems can be solved using basic iterative procedures.[3.11] Unlike the solutions to linear problems which are always unique, non-linear problem solutions are not always unique. Hence, small-step, incremental approaches may be needed to obtain good accuracy.

Another type of problem frequently encountered by designers of concrete structures is that of P–Δ effects and, more generally, large deflections. The assumption of small deflections associated with linear analysis means, in practical terms, that the geometry of the structure (and all its elements) remains essentially unchanged during the loading process. This is, of course, never strictly correct. In the event that more accurate displacement calculations are required, geometric non-linearities must be accounted for. As for material non-linearities, these can be handled using iterative, step-wise procedures.

Fortunately, advances in computer software and hardware have made it increasingly easy to perform quite complex analyses, something which not long ago was considered only the domain of researchers. For example, many structural analysis packages now allow the designer to model non-linear materials (see Table 3.1) and can also handle problems which are geometrically non-linear. These enhancements have been made it relatively easy for the engineer to undertake non-linear analysis for static loads. Even PCs can perform the calculations reasonably quickly.

On the other hand, general non-linear analysis for dynamic loading, whether the non-linearity is material or geometric in nature, is still a relatively time-consuming task and involves much computational effort. For this reason, full non-linear dynamic analysis is rarely used. However, one of the most likely instances where such an analysis would be required would be for an especially critical or sensitive structure subject to large earthquake loading. In cases where certain aspects of structural damping are the key to the structure's performance, time history analysis is also required in order to estimate the response under repeated cyclic loading such as would occur in a large earthquake. However, even for these situations the resulting maxima and minima for results of the non-linear dynamic analysis would be checked against the results of an equivalent static analysis with any large differences being scrutinized further.

3.6   Finite Elements and Their Uses in Practical Design

Much of today's design work is facilitated using standard structural frame analysis packages which include only beam and truss elements. However, finite-element analysis may be advantageous, required in fact, for structures which are highly irregular (geometrically or materially) or for which the essential structural elements consist of plates or shells acting in three dimensions.[3.12] Included in Table 3.1 is a column identifying which software packages have finite-element capability.

A hyperbolic cooling tower structure is an example of a structure which can be easily modelled in three dimensions using membrane and/or plate finite elements for either linear or non-linear analyses. Also, in cases where shear walls have many cutout sections (Fig. 3.12) or are geometrically irregular, the effective 'wide-column' model is not appropriate. In this case significant stress concentrations and irregular deformation patterns can occur which cannot be

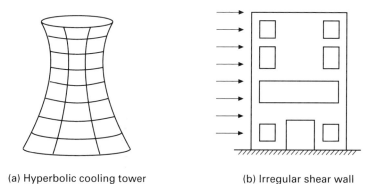

(a) Hyperbolic cooling tower   (b) Irregular shear wall

Fig. 3.12 Finite-element models of shell and shear-wall structures

detected using the simpler modelling techniques. Hence, in this case finite-element analysis may be used.

3.7   Recent and Future Developments

The recent developments may be summarized as improvements in the integration of software modules (pre-processors, analysis and design modules, and post-processors), increased design capability, and increasing user friendliness. This trend is expected to continue into the future with the result being analysis software which is well-integrated with design software. In particular, the use of non-linear analysis will become widespread. Designers will also be able to use CAD-like software to generate computer models of their designs at the preliminary design stage and develop their model through to the final design on a computer. The design calculations and construction details and drawings will also be generated by computer. As discussed briefly in Section 3.1, some software developers already claim to have this capability. Nevertheless, it is expected that this feature will soon be widely available.

Finally, another area which is receiving a great deal of attention from researchers is the use of knowledge-based, or expert system, software. In some cases, this type of computer program can be used not only to conduct structural analysis but to provide more 'personal' expert advice on topics such as modelling techniques, potential problems in the structural design etc. For example, researchers at Cornell, Lehigh and Carnegie-Mellon Universities have developed a number of knowledge-based computer programs to assist architects and engineers with the task of earthquake-resistant design of buildings. While expert system software is not yet widely used by the engineering profession, there are now several commercial programs available in Australia. For example, Windloader,[3.13] is a knowledge-based system to assist with the interpretation of the Australian Wind Loading Code, and Timber Tutor is an expert system program designed to educate and familiarize users with the Australian Timber Structures Code.

References

3.1   Clough R W, Penzien J 1993 *Dynamics of Structures* McGraw-Hill, New York

3.2   Bachmann H, Ammann W 1987 *Vibrations in Structures, Induced by Man and Machines* International Association for Bridge and Structural Engineering, Zürich

3.3   Shaeffer R E 1980 *Building Structures: Elementary Analysis and Design* Prentice Hall, New Jersey

3.4   Warner R F 1993 Non-linear design of concrete structures. *13th Australasian Conference on the Mechanics of Structures and Materials* Wollongong, pp 913–920

3.5   Chang K F 1973 Human response to motions in tall buildings. *Journal of the Structural Division, ASCE* **99**(ST6): 1259–1272

3.6   Paz M (ed) 1994 *Handbook of Multi-National Seismic Design Codes* Van Nostrand Reinhold, New York

3.7   Griffith M C, Heneker D 1993 Seismic performance of lightly reinforced concrete frame buildings. *National Earthquake Conference*, Memphis, Vol 2, pp 225–234

3.8   MacGregor J G, Lyse I (eds) 1978 *Structural Design of Tall Concrete and Masonry Buildings* Monograph on Planning and Design of Tall Buildings, Council on Tall Buildings and Urban Habitat

3.9   Coull A 1990 Analysis for structural design. *Proceedings of the 4th World Congress on Tall Buildings: 2000 and Beyond* Council on Tall Buildings and Urban Habitat, Hong Kong, pp 1031–1047

3.10   Hutchinson G L 1984 Torsional coupling effects in buildings subject to earthquake excitation. *9th Australasian Conference on the Mechanics of Structures and Materials* Sydney, pp 99–105

3.11   Zienkiewicz O C 1977 *The Finite Element Method* McGraw-Hill, London

3.12   Schlaich J M 1990 On the relevance of sophisticated modelling techniques on the design of real concrete structures. *Computer Aided Analysis and Design of Concrete Structures* Pineridge Press, Swansea, pp 1273–1281

3.13   Sharpe R, Marksjo B S, Ho F, Holmes J D 1989 *Windloader* CSIRO Division of Building, Construction and Engineering, Australia

4 Estimation and Accommodation of Column Length Changes in Tall Buildings

S. K. Ghosh

In high-rise buildings, the shortening of columns and walls caused by elastic stresses, drying shrinkage and creep can be quite significant. The absolute amount of cumulative column shortening influences the design of cladding details and the detailing of elevator rails, vertical pipes etc. The differential shortening between adjacent columns and/or walls may cause distortion of slabs, leading to impaired serviceability. A computerized procedure for prediction of elastic and inelastic column length changes is described. The procedure is applicable to concrete, steel and composite structures. Differential column length changes computed through this procedure can and should be compensated for during construction.

Length changes of exposed columns due to temperature fluctuations are transient in nature. Procedures for predicting such movements are available; estimated movements must be kept within allowable limits.

4.1 Introduction

Until the 1950s there were only a limited number of concrete buildings more than 20 storeys high. The structures of that time had heavy cladding and masonry partitions which contributed substantially to the strength and stiffness of the buildings. Also, because of the low stress levels utilized for concrete and steel, the building frame members had sizeable dimensions which resulted in substantial rigidity. The effects of frame distortions due to shrinkage, creep and temperature were secondary and could be neglected, since the capacity of the usual structure for overstress was quite high. Even wind distortions could occasionally be neglected because, although the frame was considered to provide the lateral resistance, in reality the heavy cladding resisted much of the wind.

In the late 1950s and early 1960s, the height of concrete buildings jumped from 20 to 60 storeys, all in a brief period of five to six years. The increase in height was accompanied by a sharp increase in the strength of concrete and reinforcing steel, allowing reduced cross-sections and causing a reduction in the overall rigidity. The changeover from working stress design to ultimate strength design contributed further to this trend toward smaller sections. During the same period, architects introduced the use of exposed columns that were subject to thermal movements. The above changes made it necessary for the designer of high-rise buildings to consider column length changes due to:

- elastic stresses caused by gravity loads;
- creep caused by gravity loads;

- drying shrinkage;
- temperature variations of exposed columns.

All the above distortions existed in earlier buildings; however, neglecting their effects rarely resulted in deficiencies in the serviceability of structures.

With reduced overall stiffness and with increases in height, the volume change effects became magnified and could no longer be treated as secondary considerations in design. While column length changes within a single storey may be only a few millimetres, they are cumulative. A concrete structure designed and detailed to be 300 m tall may, in reality, be only 299.7 m tall when completed, due to shortening caused by gravity loads and drying shrinkage of concrete. The cumulative differential shortening of columns also induces moments into the forcibly distorted slabs or beams and accompanying moments in the columns. The slab moments further cause load transfer to the columns which shorten less from the columns which shorten more.

In a northern US location, an average 19°C temperature differential between a typical interior column and a partially exposed exterior column would cause an average differential shortening of 0.8 mm per storey, resulting in a transient differential movement of 64 mm at the top of an 80-storey building. An elastic, creep and shrinkage strain differential of 230×10^{-6} between a heavily stressed, highly reinforced column and a more lightly stressed, lightly reinforced wall would amount to a shortening of 0.9 mm per storey, resulting in a permanent distortion of 94 mm toward the top of an 80-storey building. Such differential distortions could cause large frame moments (particularly in the slabs) and also slab tilt, affecting serviceability. The transient temperature movements must be limited to tolerable levels through modifications of column exposure, addition of insulation, or changes in the structural system. By contrast, much larger anticipated differential length changes due to elastic stresses, creep and shrinkage can be accommodated by 'cambering'* of formwork (compensation) during construction, if the magnitudes can be predicted during design.

4.2   Thermal Movements

During the mid 1960s, structural solutions were prepared and details were suggested for dealing with temperature movements of exposed columns.[4.1, 4.2, 4.3]

For temperature effects due to column exposure, a methodology was developed to determine:

- design temperatures based on geographic location and size of column;
- thermal gradients in exposed columns;
- resulting structural effects.

For design temperatures, maps of lowest winter and highest summer mean daily temperatures with frequencies of recurrence of once in 40 years were

---

* This word conventionally refers to formwork for beams and slabs that is higher at mid-span than at the edges. Here it is used to describe tilted formwork that is higher along one edge with respect to the opposite parallel edge.

prepared for members less than 300 mm thick. For thicker members up to 600 mm thick, the use of a two-day mean temperature was suggested, to account for the time lag of penetration of ambient temperature fluctuations. A graphical approach to determining isotherms, thermal gradients and average temperatures in exposed columns was developed based on the approach used by hydraulic engineers in dealing with seepage through porous soils. Having the thermal gradients, the consequent bowing and length changes of the columns could be determined, and hence the resulting structural effects including the distortions of the structure.

Along with the procedure of predicting the seasonal transient column length changes, practical means were suggested for limiting such thermal movements so that they would remain within acceptable limits. Control of thermal movements can be achieved either by regulating the amount of column exposure or by insulating the columns. If these measures cannot sufficiently reduce the anticipated distortions, then slab hinging details or a different structural configuration must be chosen.

Following the initial instances in the mid 1960s of partition distress caused by thermal movements of exposed columns, manufacturers improved the details of gypsum board partition assemblies, thus substantially increasing the deformability of partitions. Also, structural designers have since usually taken care to either control column exposure or to provide the necessary design details to avoid thermal distress likely to be caused by column exposure.

Observations in the 1960s showed that buildings up to 12–14 storeys in height with partially exposed columns did not experience distress of partitions, even in the severe Chicago climate. For buildings taller than 20 storeys with exposed columns, an analysis of thermal effects is needed to predict the magnitude of column movement. With this information, a designer can determine whether it is necessary to incorporate special details to avoid distress of non-structural elements.

4.3   Shrinkage, Creep and Elastic Movements

The presence of vertical reinforcement in a concrete column reduces the amount of shortening due to shrinkage and creep that would have taken place in the same column in the absence of any reinforcement. With the passage of time, shrinkage and creep cause load to be transferred from the concrete to the reinforcing steel. In a heavily reinforced column, all the load may eventually be transferred to the steel, with further shrinkage actually causing tension in the concrete. On the other hand, in columns with a very low percentage of reinforcement, all the steel may yield in compression. It is important to note that the overall load-carrying capacity is not affected by this load transfer; only the proportion of load carried by the two materials changes.

With respect to overall structural behaviour, cumulative column shortening causes distortions in a structure which become quite significant with increasing height. In the late 1960s, creep and shrinkage effects in buildings in the 45- to

50-storey height range were investigated and structural solutions were published.[4.4,4.5]

In recent years the height of concrete and concrete–steel composite structures has increased into the 70- to 80-storey range and beyond, making it desirable to extend the earlier solutions to cover new heights. Also, computers are invariably used these days in structural engineering offices to solve problems requiring a great deal of meticulous arithmetic calculations and 'bookkeeping'.

The structural solutions for the effects of shrinkage and creep which were developed in the 1960s required extensive longhand computations and summations because every storey-high column segment in a multistorey building is loaded in as many increments as there are storeys above. At each loading increment, each column segment has new time-dependent properties (modulus of elasticity, creep and shrinkage coefficients) as well as a new transformed section size and a new ratio of reinforcement (based on transformed section). Due to this complexity, very few structures were analysed for the effects of shrinkage and creep. Although these effects were mostly within tolerable limits for buildings in the 40- to 50-storey height range, there were some instances where the neglect caused performance problems.

In buildings 70 to 80 storeys high, it is not only the differential column length changes due to shrinkage and creep that may be significant; the differential elastic shortenings due to gravity loads may also cause unacceptable slab tilt, particularly in structures containing vertical elements of both steel and concrete. Although the total shortenings of steel and concrete are in the same range, their respective elastic components differ greatly from each other.

A procedure for determining elastic and inelastic column length changes in tall structures, developed earlier by Fintel and Kahn,[4.4,4.5] was later updated, computerized and made applicable to both concrete and steel, as well as composite structures. The rest of this chapter is devoted to this updated procedure[4.6] which separately considers elastic and creep shortenings due to gravity loads and shrinkage shortening. While elastic and creep deformations depend upon loading history, member size and reinforcement, the shrinkage component is independent of loading and depends upon member size and reinforcement. Time and material properties, of course, are important variables in all computations of shortening.

4.4   Effects of Column Shortening

The strains in the columns of low as well as ultra-high-rise buildings are similar if the stress levels are similar; however, the overall column shortening is cumulative and depends upon the height of the structure. For example, in an 80-storey steel structure, the total elastic shortening of the columns may be as high as 180–255 mm due to the high design stress levels of modern high-strength steels. By comparison, in an 80-storey concrete building, the elastic shortening of columns would amount to only about 65 mm; however, the total length change of the reinforced concrete columns may be 180–230 mm due to shrinkage and creep.

The potentially harmful effects of these large shortenings can be contained by providing details at each level that will allow the vertical structural members to deform without stressing the cladding, partitions, finishes etc. However, such details cannot eliminate the structural consequences of the relative shortening between adjacent vertical members; this shortening distorts the slab supported by the vertical members from its intended position.

Differential elastic shortenings of vertical members result from differing stress levels. Differential creep strains in concrete vertical members result from differing stress levels, loading histories, ratios of reinforcement, volume-to-surface ratios and environmental conditions. Differential shrinkage strains are independent of stress levels; they depend only upon ratios of reinforcement, volume-to-surface ratios and environmental conditions.

The shortening of columns within a single storey affects the partitions, cladding, finishes, piping etc., since these non-structural elements are not intended to carry vertical loads and are therefore not subject to shortening. On the contrary, partitions and cladding may elongate from moisture absorption, pipes from high temperature of liquid contents, cladding from solar radiation and so on. Details for attaching these elements to the structure must be planned so that their movement relative to the structure will not cause distress.

The cumulative differential shortening of columns causes the slabs to tilt with resulting rotation of partitions, as shown in Fig. 4.1. Modern dry-wall partitions can be detailed with sufficient flexibility along their peripheries and at the vertical butt joints to permit their distortion without visible distress (Fig. 4.2). Plaster and masonry partitions, which were quite common in the past, are characteristically rigid and brittle and have limited ability to undergo distortions without cracking. When a slab carrying such partitions is subject to differential support displacements, the partitions must be detailed around their peripheries to allow movement relative to the frame.

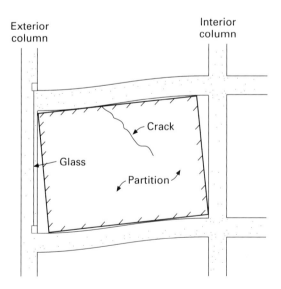

Fig. 4.1 Effect of tilted slabs

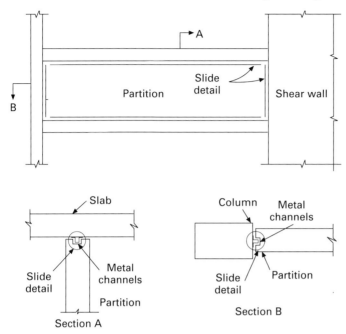

Fig. 4.2 Isolation of partitions from structural framing

4.5   Movements Related to Construction Sequence

Considering a slab at level N of a multistorey building (Fig. 4.3), each of its supports consists of N single-storey segments.

During the construction process, each of the storey-high support segments undergoes elastic shortening due to all the loads applied after placing or installation of the segment. In addition, concrete and composite columns begin to shrink from moisture loss and to creep as a result of the applied compressive forces.

The major objective of the procedure reported in this chapter is to assure a proper final position for each slab within the building. The first step is to determine the elevations of the tops of supports at the time of slab installation to assure that proper compensation can be made for installation of the slab at a predetermined position. Next to be considered are the changes of support elevations subsequent to slab installation due to elastic stresses caused by additional loads, creep caused by additional and existing loads and shrinkage.

The time of slab installation at its initial position becomes the dividing line between the following two types of support shortening.

1. Those taking place up to the time the slab is installed (pre-installation shortenings). Either analytical estimates or on-the-spot measurements of these shortenings are needed to adjust the support elevations so that the slab is installed at a predetermined position.

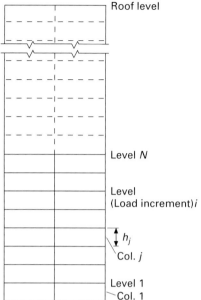

Roof level

Level N

Level
(Load increment)i

h_j

Col. j

Level 1
Col. 1

Fig. 4.3 Schematic section of a multistorey building

2. Those taking place after the slab is installed (post-installation shortenings). Analytical estimates of these shortenings are needed for corrective measures to assure that the slab will be in the desired, predetermined position after all loads have been applied and after all shrinkage and creep have taken place.

In cast-in-place reinforced concrete structures (Fig. 4.4(a)), the amount of support shortening before slab installation is of no importance, since the forms are usually levelled at the time the concrete for each floor slab is placed. However, information is needed on how much the slab will change its position after placing, from subsequent loads and subsequent volume changes. This information can then be used to tilt the formwork in the opposite direction so that in the future the slab will end up in the desired position. Depending upon circumstances, it may be decided that most of the time-dependent shortening (shrinkage and creep) be compensated for at the time of construction. In that case, at initial occupancy, the slab may have a reverse tilt that will gradually disappear. Or, it may be decided, for example, to compensate for only the shortening that is expected to take place within two years after construction. Thus, in two years the slab will be level and from that time on only the remaining shrinkage and creep will cause the slab to tilt.

In structures in which columns are fabricated to exact lengths (steel columns, Fig. 4.4(c), or light steel erection columns that are later embedded in concrete, Fig. 4.4(b)), the pre-installation support shortening is of consequence since the attachments to receive the slabs are part of the shop-fabricated columns. To assure the predetermined initial slab elevation, the pre-installation length changes of these columns need to be known and compensated for. The post-installation support shortenings (elastic and inelastic) must also be considered.

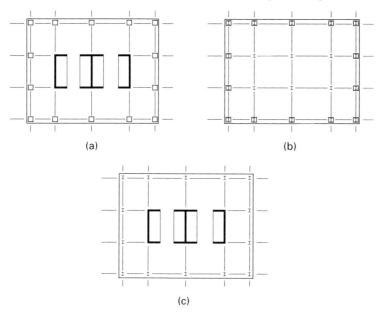

Fig. 4.4 Structural systems for tall buildings: (a) concrete frame–shear-wall interactive system; (b) interior steel framing with peripheral concrete framing; (c) interior concrete shear walls with exterior steel framing

4.6   Relative Movements between Adjacent Elements in Various Structural Systems

4.6.1   Reinforced Concrete Structures

As mentioned, only the differential movements that will occur between the supports of a slab after it has been placed are of concern, since they will change the position of the slab.

The total shortening is rarely of practical interest. Even in cases of elevator rails attached to the shaft structure or vertical pipes, only the shortening that will occur after their attachment is of concern.

Differential movements in concrete buildings (Fig. 4.4(a)) are primarily those between isolated columns and shear walls due to different ratios of vertical reinforcement, different stress levels and different volume-to-surface ratios. Differential movements between neighbouring columns may also occur. Differential shortenings are of particular concern where the distance between differentially shortening elements is small, causing significant tilt of the beam or the slab. Strain differentials of up to 200×10^{-6} have been observed in cases where all the contributing factors increase the differential strains in the same direction. Such a strain differential amounts to a shortening of 0.73 mm in a 3.7 m storey and in a 50-storey building would amount to 36.5 mm of total differential shortening. Obviously, such differential length changes need to be compensated for – at least the part that will occur within a certain initial period (say, two years, during which the majority of shrinkage and creep will have taken place), thus leaving only a minor portion of the total shrinkage and creep movement to cause slab or beam tilt.

4.6.2   Composite Structures

Two types of composite (steel and concrete) structural systems are currently in popular use; both utilize structural slabs consisting of steel beams supporting corrugated decks topped with concrete. One type has a reinforced concrete core with exterior structural steel columns (Fig. 4.4(c)); the other consists of a peripheral reinforced concrete column–beam system with structural steel interior columns (Fig. 4.4(b)). A variant of the latter system that facilitates a more efficient construction procedure is a basic steel structure with light peripheral columns (called erection columns) that are later encased in concrete. A third type of composite structural system consisting of a heavy structural steel core and exterior composite columns (steel erection columns encased in reinforced concrete) has also been used in recent times. For the purposes of this chapter, there is no essential difference between this system and the one shown in Fig. 4.4(b). Thus, the third system is not considered separately.

In composite structures with erection columns, the progress of peripheral concrete column casting follows behind the top lift of steel erection by a certain number of storeys, say nine. This nine-storey steel skeleton (running ahead of the peripheral concrete columns) consists of (a) three storeys of steel columns and girders, (b) three storeys of steel columns and girders plus steel deck and (c) three storeys of steel columns and girders plus steel deck plus concrete topping. As a new lift of steel columns and girders is added at the top, simultaneously three storeys below, a steel deck is added. Three storeys farther down, concrete topping is added to a steel deck; and three storeys still farther down, a lift of concrete peripheral columns is cast. Thus, there is always a nine-storey steel structure above the storey on which concrete columns have been cast (except at the very top where concrete lifts gradually approach the roof level after steel lifts have been erected for the entire building).

Composite Structures with Erection Columns
The erection columns are designed to carry only nine to 12 storeys of construction loads consisting of three to four storeys each of the stages (a), (b) and (c) described above. These columns are then embedded into the peripheral concrete columns and become part of their vertical reinforcement.

At the time of embedment, an erection column is stressed to about 60 per cent of the yield strength of the column steel and has experienced a corresponding elastic shortening. Thus, the storey-high lifts of the erection columns should be made longer to compensate for the pre-embedment differential shortening between them and the interior columns that are subjected to significantly lower stress levels. The loads carried by erection columns before their embedment are carried directly in steel-to-steel bearing, starting from the foundation-level base plate. At the time of casting, each lift of a composite column is subjected to the column dead load and one added full-floor load consisting of a concrete topping three storeys above, a steel deck three storeys farther up and a new lift of steel columns and girders three storeys still farther up. During construction of the top nine storeys of the building, the additional loads imposed on new lifts of composite columns are progressively smaller than full-floor loads, as the roof is approached. For example, when a column just below the roof level is cast, the only additional load on it is that of the column itself.

With the progress of time after construction, the elastic and inelastic strains of a composite column are added to the initial strains that the erection column had prior to embedment. At some point, the embedded erection column may begin to yield; further loads will then be carried by the concrete and the vertical reinforcement not yet yielded. It should be noted that such load shifting between reinforcing steel and concrete does not affect the overall load-carrying capacity of a column; only the proportion of load carried by the steel and the concrete changes.

Composite Structures with Concrete Cores

The composite-type structure consisting of a reinforced concrete central core and peripheral steel columns might have the core slipformed ahead of the steel structure; or the core might be built using jump forms proceeding simultaneously with the steel structure; or the core might follow behind steel erection by nine to 12 storeys, as described in the previous section. From the point of view of relative shortening between the core and the steel columns, all three cases are quite different. In the case of the slipformed core, at the time a steel storey is erected, some of the creep and shrinkage of the core will already have taken place. Thus, the post-slab-installation portion of shrinkage and creep is smaller. In cases where the core construction proceeds simultaneously with or lags behind the steel structure, all of the shrinkage and creep of the core contributes to the differential shortening relative to the steel columns.

4.7   Estimates of Elastic, Shrinkage and Creep Strains in Columns

To carry out an analysis for elastic, shrinkage and creep strains in reinforced concrete columns, information is required on the modulus of elasticity and the shrinkage and creep characteristics of the concrete mixes to be used in the structure being considered. Discussion of these properties, abbreviated from Ref. 4.6, follows.

4.7.1   Modulus of Elasticity

The effect on modulus of elasticity of the age of concrete at the time of loading is taken into account implicitly when the strength of the concrete at the time of loading, rather than at 28 days, is inserted into the modulus of elasticity expression given in the ACI Code:[4.7]

$$E_{ct} = 0.043w^{1.5}\sqrt{f'_{ct}}$$ [4.1]

where E_{ct} is the time-dependent modulus of elasticity of concrete in MPa, w is the unit weight of concrete in kg m^{-3} and f'_{ct} is the time-dependent concrete compressive strength in MPa.

The American Concrete Institute (ACI) Committee 209[4.8] has recommended the following expression for the time-dependent strength of moist-cured (as distinct from steam-cured) concrete using Type I cement:

$$f'_{ct} = \frac{t}{4.0 + 0.85t} f'_c$$ [4.2]

where t is the time in days from placing to loading and f'_c is the 28-day compressive strength of concrete.

A version of Eq. 4.1, applicable to concretes with compression strengths of up to 80 MPa (the ACI Code formula for modulus of elasticity was originally derived for concretes with compression strengths of no more than 40 MPa), is now available in Refs 4.9 and 4.10 (see also Eq. 7.6 in Chapter 7).

The validity of Eq. 4.1 at very early concrete ages was felt open to question. A rounded off version of Eq. 4.1 for normal-weight concrete was compared in Ref. 4.6 with the experimental information available on the modulus of elasticity of concrete at early ages.[4.11, 4.12] In the range of practical application, the ACI equation relating modulus of elasticity and strength was found to agree quite favourably with the trend of experimental results.

4.7.2   Shrinkage of Unreinforced (Plain) Concrete

Shrinkage of concrete is caused by evaporation of moisture from the surface. The rate of shrinkage is high at early ages and decreases with an increase in age until the curve becomes asymptotic to the final value of shrinkage. The rate and amount of evaporation and consequently of shrinkage depend greatly upon relative humidity of the environment, size of the member and mix proportions of the concrete. In a dry atmosphere, moderate-size members (600 mm diameter) may undergo up to half of their ultimate shrinkage within two to four months, while identical members kept in water may exhibit growth instead of shrinkage. In moderate-size members, the inside relative humidity has been measured at 80% after four years of storage in a laboratory at 50% relative humidity.

Basic Value of Shrinkage

Let ε_s denote the ultimate shrinkage of 150 mm-diameter standard cylinders (volume-to-surface or $v:s$ ratio $= 38$ mm) moist-cured for seven days and then exposed to 40% ambient relative humidity. If concrete has been cured for less than seven days, multiply ε_s by a factor linearly varying from 1.2 for one day of curing to 1.0 for seven days of curing.[4.8]

Attempts have been made in the past to correlate ε_s with parameters such as concrete strength. In view of available experimental data,[4.13] it appears that no such correlation may in fact exist. The only possible correlation is probably that between ε_s and the water content of a concrete mix.[4.14] In the absence of specific shrinkage data for concretes to be used in a particular structure, the value of ε_s may be taken as between 500×10^{-6} (low value) and 800×10^{-6} (high value). The latter value has been recommended by ACI Committee 209.[4.8]

Effect of Member Size

Since evaporation occurs from the surface of members, the volume-to-surface ratio of a member has a pronounced effect on the amount of its shrinkage. The amount of shrinkage decreases as the size of specimen increases.

For shrinkage of members having volume-to-surface ratios different from 1.5, ε_s must be multiplied by the following factor:

$$SH_{v:s} = \frac{0.015(v:s) + 0.944}{0.070(v:s) + 0.734} \qquad [4.3]$$

where $v:s$ is the volume-to-surface ratio in mm.

As indicated in Ref. 4.6, Eq. 4.3 is based on laboratory data[4.15] and European recommendations.[4.16,4.17]

Much of the shrinkage data available in the literature was obtained from tests on prisms of a 75×75 mm section ($v:s = 19$ mm). According to Eq. 4.3, the size coefficient for prisms of that size is 1.12. Thus, shrinkage measured on a prism of a 75×75 mm section must be divided by 1.12, before the size coefficient given by Eq. 4.3 is applied to it. It should be cautioned, however, that as the specimen size becomes smaller, the extrapolation to full-size members become less accurate.

Effect of Relative Humidity

The rate and amount of shrinkage greatly depend upon the relative humidity of the environment. If ambient relative humidity is substantially greater than 40 per cent, ε_s must be multiplied by

$$\begin{aligned} SH_H &= 1.40 - 0.010H \text{ for } 40 \le H \le 80 \\ &= 3.00 - 0.030H \text{ for } 80 \le H \le 100 \end{aligned} \qquad [4.4]$$

where H is the percentage relative humidity. Average annual values of H should probably be used. Maps giving average annual relative humidities for locations around the United States are available.[4.6] However, if locally measured humidity data are available, they are likely to be more accurate than the information such as that included in Ref. 4.6 and should be used in conjunction with Eq. 4.4.

Equation 4.4 is based on ACI Committee 209 recommendations.[4.8] A comparison with European recommendations[4.16] is shown in Ref. 4.6.

If shrinkage specimens are stored under jobsite conditions rather than under standard laboratory conditions, the correction for humidity, as given by Eq. 4.4, should be neglected.

Progress of Shrinkage with Time

Hansen and Mattock[4.15] established that the size of a member influences not only the final value of shrinkage but also the rate of shrinkage, which appears to be only logical. Their expression giving the progress of shrinkage with time is

$$SH_t = \frac{\varepsilon_{st}}{\varepsilon_s} = \frac{t}{26.0e^{0.36(v:s)} + t} \qquad [4.5]$$

where ε_{st} and ε_s are shrinkage strains up to time t and time infinity, respectively; and t is measured from the end of moist curing.

Equation 4.5 is compared in Fig. 4.5 with the progress of shrinkage curve from Ref. 4.16. Also shown in Fig. 4.5 is a comparison of Eq. 4.5 with the progress of shrinkage relationship recommended by ACI Committee 209.[4.8] It should be noted that both the ACI–Cement and Concrete Association (C&CA)[4.16] and the ACI Committee 209[4.8] relationships are independent of the volume-to-surface ratio.

4.7.3   Creep of Unreinforced (Plain) Concrete

Creep is a time-dependent increment of the strain of a stressed element that continues for many years. The basic phenomenon of creep is not yet conclusively explained. During the initial period following the loading of a structural member, the rate of creep is significant. The rate diminishes as time progresses until it eventually approaches zero.

Creep consists of two components:

1. Basic (or true) creep occurring under conditions of hygral equilibrium, which means that no moisture movement occurs to or from the ambient medium. In the laboratory, basic creep can be reproduced by sealing a specimen in copper foil or by keeping it in a fog room.

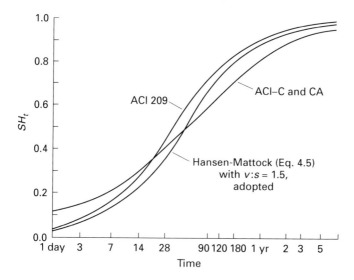

Fig. 4.5 Progress of shrinkage with time

2. Drying creep resulting from an exchange of moisture between the stressed member and its environment. Drying creep has its effect only during the initial period under load.

Creep of concrete is very nearly a linear function of stress up to stresses that are about 40 per cent of the ultimate strength. This includes all practical ranges of stresses in columns and walls. Beyond this level, creep becomes a non-linear function of stress.

For structural engineering practice it is convenient to consider specific creep, which is defined as the ultimate creep strain per unit of sustained stress.

Value of Specific Creep

Specific creep values can be obtained by extrapolation of results from a number of laboratory tests performed on samples prepared in advance from the actual mix to be used in a structure. It is obvious that sufficient time for such tests must be allowed prior to the start of construction, since reliability of the prediction improves with the length of time over which creep is actually measured.

A way of predicting basic specific creep (excluding drying creep), without testing, from the modulus of elasticity of concrete at the time of loading was proposed by Hickey[4.18] on the basis of long-term creep studies at the Bureau of Reclamation in Denver. Hickey's proposal was adopted by Fintel and Khan.[4.4, 4.5]

Let ε_c denote the specific creep (basic plus drying) of 150 mm-diameter standard cylinders ($v:s = 38$ mm) exposed to 40 per cent relative humidity following about seven days of moist-curing and loaded at the age of 28 days. In the absence of specific creep data for concretes to be used in a particular structure, the following likely values of ε_c are recommended:

$$\varepsilon_c = 0.000435/f_c' \text{ (low value) to } 0.000725/f_c' \text{ (high value)} \qquad [4.6]$$

where ε_c is in $mm^{-1}\ MPa^{-1}$ when f_c' is in MPa. The lower end of the proposed range is in accord with specific creep values suggested by Neville.[4.19] The upper end agrees with laboratory data obtained by testing concretes used in Water Tower Place[4.13] in Chicago, Illinois.

Effect of Age of Concrete at Loading

For a given mix of concrete the amount of creep depends not only on the stress level, but also to a great extent on the age of the concrete at the time of loading. Figure 4.6 shows the relationship between creep and age at loading as developed by Comité Europeen du Béton (CEB), using available information from many tests.[4.17] The coefficient CR_{LA} relates the creep for any age at loading to the creep of a specimen loaded at the age of 28 days. The 28-day creep is used as a basis of comparison, the corresponding CR_{LA} being equal to 1.0.

Figure 4.6 also depicts the following suggested relationship between creep and age at loading:

$$CR_{LA} = 2.3t_{LA}^{-0.25} \qquad [4.7]$$

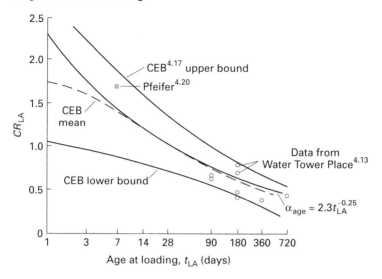

Fig. 4.6 Creep versus age of concrete at time of loading

where t_{LA} is the age of concrete at the time of loading, in days. The form of Eq. 4.7 is as suggested by ACI Committee 209.[4.8] Equation 4.7 gives better correlation with the CEB mean curve than the corresponding equation suggested by Committee 209. Figure 4.6 also shows comparison with a few experimental results.[4.13, 4.20] According to Eq. 4.7, the creep of concrete loaded at seven days of age is 41 per cent higher than that of concrete loaded at 28 days.

Effect of Member Size

Creep is less sensitive to member size than shrinkage, since only the drying-creep component of the total creep is affected by the size and shape of members, whereas basic creep is independent of size and shape.

For members with volume-to-surface ratios different from 38 mm, ε_c should be multiplied by

$$CR_{v:s} = \frac{0.017(v:s) + 0.934}{0.039(v:s) + 0.85} \qquad [4.8]$$

where $v:s$ is the volume-to-surface ratio in mm.

As indicated in Ref. 4.6, Eq. 4.8 is based on laboratory data[4.15] and European recommendations.[4.16, 4.17] Much of the creep data available in the literature was obtained by testing 150 mm-diameter standard cylinders wrapped in foil. The wrapped specimens simulate very large columns. Equation 4.8 yields a value of $CR_{v:s}$ equal to 0.49 for $v:s = 100$. Thus, it is suggested that creep data obtained from sealed specimen tests should be multiplied by 2 ($1/0.49 \cong 2$) before the modification factor given by Eq. 4.8 is applied to such data.

Effect of Relative Humidity

For an ambient relative humidity greater than 40 per cent, ε_c should be multiplied by the following factor, as suggested by ACI Committee 209:[4.8]

$$CR_H = 1.40 - 0.01H \qquad [4.9]$$

where H is the percentage relative humidity. Again, it is suggested that the average annual value of H should be used.

Progress of Creep with Time

The progress of creep relationship recommended by ACI Committee 209[4.8] is given by the following expression:

$$CR_t = \frac{\varepsilon_{ct}}{\varepsilon_c} = \frac{t^{0.6}}{10 + t^{0.6}} \qquad [4.10]$$

where ε_{ct} is creep strain per unit stress up to time t, and t is measured from the time of loading.

The above relationship is plotted in Fig. 4.7 where it is seen to compare well with the creep versus time curve suggested in European recommendations.[4.16] Equation 4.10 is recommended by the author.

4.7.4   Residual Shrinkage and Creep of Reinforced Concrete

In a reinforced concrete column, creep and shrinkage of the concrete are restrained by the reinforcement. Tests have shown that when reinforced concrete columns are subjected to sustained loads, there is a tendency for stress to be gradually transferred to the steel, with a simultaneous decrease in the concrete stress. Long-term tests by Troxell and others[4.21] showed that in columns with low percentages of reinforcement, the stress in the steel increased until yielding, while

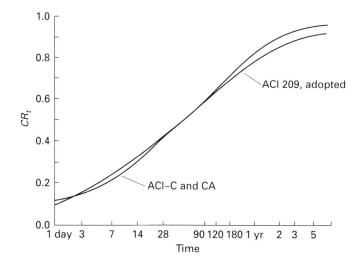

Fig. 4.7 Progress of creep with time

in highly reinforced columns, after the entire load had been transferred to the steel, further shrinkage actually caused some tensile stresses in the concrete. It should, however, be noted that despite the redistribution of load between concrete and steel, the ultimate load capacity of the column remains unchanged.

The residual creep strain of a reinforced concrete column segment can be calculated by the following formula first proposed by Dischinger in 1937[4.22] and used earlier by Fintel and Kahn:[4.4, 4.5]

$$\varepsilon_c^R = \varepsilon_c CR_R = \frac{\varepsilon_c}{\rho \varepsilon_c^* E_s} \left(1 - e^{\frac{-\rho m}{1+\rho m} \varepsilon_c^* E_{ct}} \right) \qquad [4.11]$$

where ε_c^R is the total residual (ultimate) creep strain per unit stress in reinforced concrete; CR_R is the residual creep factor; ρ is the reinforcement ratio of the cross-section of the column segment; ε_c^* is the specific creep of plain concrete, adjusted for age at loading, volume-to-surface ratio and humidity; m is the time-dependent modular ratio E_s/E_{ct}; and E_s is the modulus of elasticity of steel.

Since ε_c^* (which includes adjustment for age at loading), E_{ct} and m are all time-dependent, the factor CR_R, as calculated by Eq 4.11, will be different for each load increment applied to a column segment. As to shrinkage, it is suggested that for a column segment subjected to k load increments, residual shrinkage strains be calculated as follows:

$$\varepsilon_s^R = \varepsilon_s SH_R = \varepsilon_s \frac{1}{k} \sum_{i=1}^{k} CR_{R,i} \qquad [4.12]$$

where ε_s^R is the total residual (ultimate) shrinkage strain of reinforced concrete, $CR_{R,i}$ is the residual creep factor corresponding to the ith load increment and SH_R is the average residual factor for shrinkage, which is a load-independent phenomenon.

4.8   Calculation of Elastic, Shrinkage and Creep Shortening of Columns

Figure 4.3 shows a schematic section of a multistorey building with reinforced concrete or composite columns, with the slabs up to level N already cast. The slabs above level N will be cast as construction proceeds. This section presents mathematical expressions for the cumulative elastic, shrinkage and creep short-ening of all segments of a column up to level N (called solution-floor level in the remainder of this chapter). In other words, expressions are given for the vertical displacement of one support of the slab at level N.

Several possible load stages are considered. The initial loads are those that start acting immediately upon construction. These come on the structure in as many increments as there are floors. One set of subsequent loads may be those due to the installation of cladding and partitions. Installation of such items would nor-mally proceed storey by storey, so that the loads would come on in as many increments as there are storeys. The final set of loads may be live loads that start acting as the building is occupied. Occupation may proceed storey by storey or in some other sequence. The following expressions allow specification of the time of application of each stage of each floor load separately. Each type of shortening caused by the initial loads, as discussed previously, is computed separately up to and subsequent to the casting of the slab at level N.

The postulated and confirmed principle of superposition of creep states the following.

Strains produced in concrete at any time by a stress increment are independent of the effects of any stress applied either earlier or later. The stress increment may be either positive or negative, but stresses that approach the ultimate strength are excluded.

Thus, each load increment causes a creep strain corresponding to the strength-to-stress ratio at the time of its application, as if it were the only loading to which the column were subjected. This principle of superposition is applied to determine the total creep strains in a column subjected to a number of load increments by totalling the creep strains caused by each of the incremental loadings.

Elastic shortening of steel columns can be computed in exactly the same way as that of reinforced concrete or composite columns, except that the computation is somewhat simpler due to the absence of any effect of age on strength and due to the absence of shrinkage and creep.

4.8.1   Elastic Shortening (Denoted by Superscript e)

Due to Initial Loads (Denoted by Subscript 1)
Up to casting of solution-floor level (denoted by subscript p):

$$\Delta^e_{1,p} = \sum_{j=1}^{N} \sum_{i=j}^{N} \frac{P_i h_j}{A_{t,ij} E_{ct,ij}} \tag{4.13}$$

with

$$E_{ct,ij} = 0.043 w^{1.5} \sqrt{f'_{ct,ij}} \text{ (from Eq. 4.1)} \tag{4.13a}$$

$$f'_{ct,ij} = \frac{f'_{c,j}(t_i - t_j)}{4.0 + 0.85(t_i - t_j)} \text{ (from Eq. 4.2)} \tag{4.13a'}$$

and

$$A_{t,ij} = A_{gj} + A_{s,j}(m_{ij} - 1) \tag{4.13b}$$

$$m_{ij} = E_s/E_{ct,ij} \tag{4.13b'}$$

where

$i =$ a particular floor level or load increment;
$j =$ a particular column;
$P =$ applied load;
$h =$ floor height;
$A_t =$ time-dependent transformed area of column cross-section;

E_{ct} = time-dependent modulus of elasticity of concrete;
w = unit weight of concrete;
f'_{ct} = time-dependent cylinder strength of concrete;
t = time of casting or load application (starting from casting of foundation);
A_g = gross area of column cross-section;
A_s = total area of reinforcing steel in column cross-section;
m = time-dependent modular ratio;
E_s = modulus of elasticity of steel.

Subsequent to casting of solution-floor level (denoted by subscript s):

$$\Delta^e_{1,s} = \sum_{j=1}^{N} \sum_{i=N+1}^{n} \frac{P_i h_j}{A_{t,ij} E_{ct,ij}} \qquad [4.14]$$

where n is the total number of floors.

Due to Subsequent Load Applications(s) (Denoted by Subscripts 2, 3 etc.)

$$\Delta^e_2 = \sum_{j=1}^{N} \sum_{k=j}^{n} \frac{P_k h_j}{A_{t,kj} E_{ct,kj}} \qquad [4.15]$$

with

$$E_{ct,kj} = 0.043 w^{1.5} \sqrt{f'_{ct,kj}} \text{ (from Eq. 4.1)} \qquad [4.15a]$$

$$f'_{ct,kj} = \frac{f'_{c,j}(t_k - t_j)}{4.0 + 0.85(t_k - t_j)} \text{ (from Eq. 4.2)} \qquad [4.15a']$$

where k is a particular floor level or load increment.

$$\Delta^e_3 = \sum_{j=1}^{N} \sum_{l=j}^{n} \frac{P_l h_j}{A_{t,lj} E_{ct,lj}} \text{ and so on} \qquad [4.15']$$

where l is a particular floor level or load increment.

4.8.2   Shrinkage Shortening (Denoted by Superscript s)

Up to casting of solution-floor level (denoted by subscript p):

$$\Delta^s_p = \sum_{j=1}^{N} h_j \varepsilon_{s,j} SH_{v:s,j} SH_H SH_{t,j} SH_{R,j} \qquad [4.16]$$

with

$$SH_{v:s,j} = \frac{0.015(v:s)_j + 0.944}{0.070(v:s)_j + 0.734} \text{ (from Eq. 4.3)} \qquad [4.16a]$$

and

$$SH_{t,j} = \frac{t_N - t_j - t_j'}{26.0e^{0.36(v:s)_j} + (t_N - t_j - t_j')} \quad \text{(from Eq. 4.5)} \qquad [4.16b]$$

where t_j' is the period of moist-curing of column j, SH_H is from Eq. 4.4 and $SH_{R,j}$ (see Eq. 4.12) is defined as follows:

$$SH_{R,j} = \frac{\sum_{i=j}^{n} CR_{R,ij}}{n - j + 1} \qquad [4.16c]$$

$CR_{R,ij}$ is given by Eq. 4.18d.

Subsequent to casting of solution-floor level (denoted by subscript s):

$$\Delta_s^s = \sum_{j=1}^{N} h_j \varepsilon_{s,j} SH_{v:s,j} SH_H (1 - SH_{t,j}) SH_{R,j} \qquad [4.17]$$

4.8.3   Creep Shortening (Denoted by Superscript c)

Due to Initial Loads (Denoted by Subscript 1)
Up to casting of solution-floor level (denoted by subscript p):

$$\Delta_{1,p}^c = \sum_{j=1}^{N} \sum_{i=j}^{n} \frac{P_i CR_{LA,ij}}{A_{t,ij}} \varepsilon_{cj} h_j CR_{v:s,j} CR_H CR_{t,j} CR_{R,ij} \qquad [4.18]$$

where

$$CR_{LA,ij} = 2.3(t_i - t_j)^{-0.25} \quad \text{(from Eq. 4.7)} \qquad [4.18a]$$

$A_{t,ij}$ has been defined by Eqs 4.13a, 4.13a', 4.13b and 4.13b'.

$$CR_{v:s,j} = \frac{0.017(v:s)_j + 0.934}{0.039(v:s)_j + 0.85} \quad \text{(from Eq. 4.8)} \qquad [4.18b]$$

CR_H is given by Eq. 4.9

$$CR_{t,j} = \frac{(t_N - t_j)^{0.6}}{10.0 + (t_n - t_j)^{0.6}} \quad \text{if } t_N > t_j \qquad [4.18c]$$

$$= 0 \quad \text{if } t_N \leq t_j \text{ (from Eq. 4.10)}$$

and

$$CR_{R,ij} = \frac{1 - e^{\frac{-\rho_j m_{ij}}{1 + \rho_j m_{ij}} \varepsilon_{c,ij}^* E_{ct,ij}}}{\rho_j \varepsilon_{c,ij}^* E_s} \quad \text{(from Eq. 4.11)} \qquad [4.18d]$$

$$\varepsilon_{c,ij}^* = \varepsilon_{c,ij} CR_{LA,ij} CR_{v:s,j} CR_H \qquad [4.18d']$$

$$\rho_j = A_{s,j}/A_{g,j} \qquad [4.18d'']$$

Subsequent to casting of solution-floor level (denoted by subscript s):

$$\Delta_{1,s}^{c} = \sum_{j=1}^{N} \sum_{i=j}^{n} \frac{P_i CR_{LA,ij}}{A_{t,ij}} \varepsilon_{c,j} h_j CR_{v:s,j} CR_H (1 - CR_{t,j}) CR_{R,ij} \qquad [4.19]$$

Due to Subsequent Load Application(s) (Denoted by Subscripts 2, 3 etc.)

$$\Delta_{2}^{c} = \sum_{j=1}^{N} \sum_{k=j}^{n} \frac{P_k CR_{LA,kj}}{A_{t,kj}} \varepsilon_{c,j} h_j CR_{v:s,j} CR_H CR_{R,kj} \qquad [4.20]$$

$$\Delta_{3}^{c} = \sum_{j=1}^{N} \sum_{l=j}^{n} \frac{P_l CR_{LA,lj}}{A_{t,lj}} \varepsilon_{c,j} h_j CR_{v:s,j} CR_H CR_{R,lj} \qquad [4.20']$$

4.9   Examples of Column Shortening Analysis

Using the methodology described in the previous sections, examples of analyses for differential column length changes in tall buildings featuring two different structural systems are presented below.

4.9.1   Eighty-Storey Composite Structure

Figure 4.8 shows a quarter of the plan of an 80-storey composite building with a peripheral beam–column system of reinforced concrete to provide lateral ri-

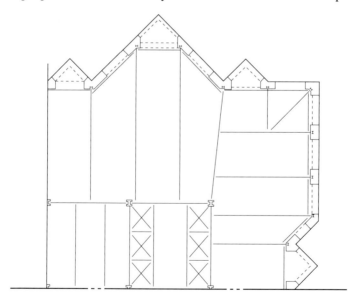

Fig. 4.8 Typical floor plan of 80-storey steel-concrete composite building

gidity and interior structural steel columns supporting a slab system of structural steel beams, a corrugated deck and a concrete topping. The large peripheral columns have structural steel erection columns embedded in them. Table 4.1 gives the loads and the section properties for a typical exterior and a typical interior column throughout the height of the building. The computed components of shortening due to elastic stresses, shrinkage and creep for the exterior composite column and the interior steel column are shown in Fig. 4.9.

Curve a in Fig. 4.9(a) (showing the elastic shortening of the interior steel column) indicates that the vertical column displacements at the various floor levels up to the time of slab installation at those levels increase up the height of the building, because the loads from each added floor shorten all the column segments below that level.

Subsequent to slab installation at the various levels, the vertical column displacements at those levels initially increase, but then decrease with increasing height (curve b, Fig. 4.9(a)). This is because loads from fewer and fewer storeys contribute to post-slab-installation shortenings as construction progresses toward the top of the building. At the roof level, only the mechanical bulkheads above contribute to column shortening subsequent to the installation of the roof slab.

Curves 1 and 5 in Fig. 4.9(b) (showing the elastic shortening of the exterior composite column) exhibit the same trends as curves a and b, respectively, of Fig. 4.9(a). The shortenings of exterior columns are based on assumed moderate values of shrinkage and creep coefficients (Table 4.1). As mentioned, light erection columns are embedded in the peripheral concrete columns. Their elastic shortenings prior to embedment become part of the total differential shortenings between exterior and interior columns (Fig. 4.9(b)).

Figure 4.9(c) shows the differential shortenings at the various floor levels between the interior steel and the exterior composite columns considered in Figs 4.9(a) and (b). The differential shortenings up to and subsequent to the installation of slabs at the various floor levels have been added for compensation purposes. The values shown on the right of Fig. 4.9(c) are needed to detail the columns for fabrication, so that after all loads have been applied and shrinkage and creep have taken place, the slabs will be horizontal. The largest differential shortening over the height of the structure is 67 mm; this maximum occurs at the 60th floor level.

Figures 4.10(a) and (b) are similar to Figs 4.9(b) and (c), respectively, and show shortenings of the exterior composite column for assumed high values of shrinkage and creep (Table 4.1). The maximum differential shortening over the height of the structure is now 109 mm occurring at the 70th floor level; obviously, corrective measures are required during construction to avoid tilted slabs. For this particular building, it is suggested that at every 10th storey the interior column lift be shortened as shown on the compensation curve.

Table 4.1 Properties of exterior composite column and interior steel column in an 80-storey composite building (speed of construction: seven days per floor; storey height: 5.5 m bottom storey; 4 m, all other storeys)

Exterior composite column

Floor levels	Concrete strength (MPa)	Gross col. area (×10⁶ mm²)	Steel area* (×10³ mm²)	Floor load (kN)	Subsequent floor load† (kN)	Volume-to-surface ratio (mm)	Ultimate shrinkage (×10⁻⁶)	Specific creep (×10⁻⁶ per kPa)
1–9	48	1.82	48.7	283.8	70.7	322.6		0.029‡
10–19	48	1.56	44.8	244.2	60.9	269.2		0.058§
20–29	41	1.56	28.6	244.2	60.9	269.2		0.036‡
30–39	41	1.28	25.8	218.8	54.7	228.6	600‡	0.065§
40–49	41	1.28	25.8	198.8	49.8	228.6	800§	
50–59	34	1.28	25.8	198.8	49.8	228.6		
60–69	34	1.28	19.4	202.4	50.7	228.6		0.044‡
70–76	34	1.28	19.4	212.6	52.9	228.6		0.073§
77	34	1.28	19.4	719.2	179.7	228.6		

* Includes erection column area.
† Assumed to start acting 300 days after casting of column.
‡ Low value.
§ High value.

Table 4.1 (*contd.*)
Interior steel column

Floor levels	Gross area ($\times 10^5$ mm^2)	Floor load (kN)	Floor levels	Gross area ($\times 10^5$ mm^2)	Floor load (kN)
1	2.78	361.6	56–57	0.86	334.0
2–7	2.65	361.6	58–59	0.81	334.0
8–9	2.53	361.6	60–61	0.75	355.8
10–13	2.53	350.5	62–63	0.70	355.8
14–19	2.30	350.5	64–65	0.65	355.8
20–25	2.30	346.5	66–67	0.54	355.8
26–29	1.51	346.5	68–69	0.49	355.8
30–37	1.51	366.1	70–71	0.44	328.7
38–39	1.39	366.1	72–73	0.40	328.7
40–45	1.26	338.5	74–75	0.33	328.7
46–47	1.15	338.5	76–77	0.29	328.7
48–49	1.05	338.5	78	0.23	328.7
50–51	1.05	334.0	79	0.23	2895.6
52–55	0.95	334.0			

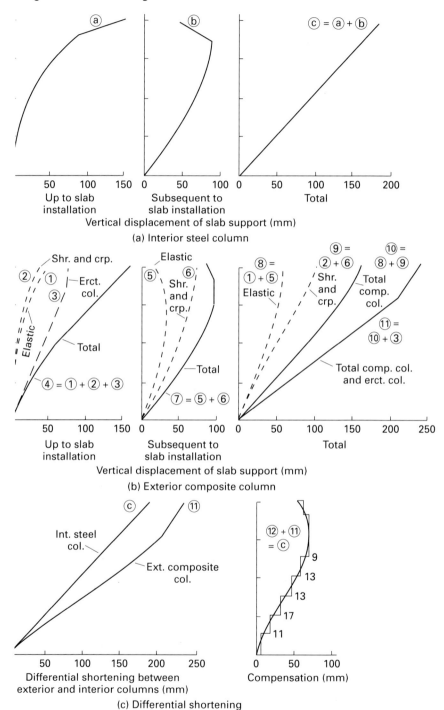

Fig. 4.9 Column length changes in 80-storey building with composite steel–concrete structural system, assuming low shrinkage and creep coefficients

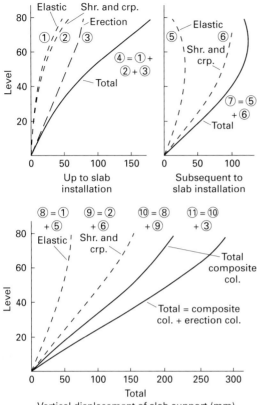

(a) Exterior composite column

(b) Differential shortening

Fig. 4.10 Column length changes in 80-storey building with composite steel–concrete structural system, assuming high shrinkage and creep coefficients

4.9.2   Seventy-Storey Reinforced Concrete Frame–Shear-wall Building

Figure 4.11 shows the plan of a 70-storey reinforced concrete frame–shear-wall building. In Table 4.2, concrete strengths, loads, amounts of reinforcement and other properties for a typical peripheral column and a typical interior shear-wall segment are given. As can be judged from the plan, the potential differential shortening between the peripheral columns and the central shear-wall core may cause tilting of the slabs and should be investigated.

The components of elastic, shrinkage and creep shortenings of the peripheral column and the interior shear-wall segment are shown in Fig. 4.12 for relatively high assumed values of shrinkage and creep (Table 4.2). The deformations that occur before the casting of a slab are of no consequence, since the formwork for each slab is usually installed horizontally; thus, the pre-slab-installation differentials are automatically compensated for. Only the post-slab-installation deformations (right-hand side of Fig. 4.12(b)) may need compensation, if the predicted amount is more than can be tolerated.

To determine the effect of a range of material properties, the shortenings of the exterior column and the interior shear-wall segment also were computed for relatively low assumed values of shrinkage and creep (Table 4.2); these are presented in Fig. 4.13. The right-hand side of Fig. 4.13(b) shows the differential shortenings (after slab casting) for low values of shrinkage and creep.

As can be seen, the maximum differential shortenings for the 70-storey building are at about the 40th floor level, ranging between 22 mm and 27 mm for low and high values of shrinkage and creep, respectively.

It is quite simple to compensate for differential shortening in concrete structures by raising the forms along the peripheral columns.

Whether to compensate for a maximum differential shortening of only 25 mm is a question that should be decided by comparing the disadvantages of the tilted slab against the cost and inconvenience of cambering the slabs during construction.

4.9.3   Thirty-Six-Storey Reinforced Concrete Frame–Shear-Wall Building*

An analysis for column shortening was applied to the 36-storey (excluding four basement levels) reinforced concrete frame–shear-wall building for the Kwang Ju City Bank in Seoul, Korea. The typical floor plan is shown in Fig. 4.14. The concrete strengths, loads, amounts of reinforcement and other properties for a typical column C2, throughout the height of the building, are given in Table 4.3.

The computed elastic and inelastic components of shortening and the total shortening (pre- and post-slab-installation deformations lumped together) for columns C1, C2, C3 and C4 are shown in Figs 4.15(a), (b), (c) and (d), re-

---

* This case study has been contributed by Professor Sung-Woo Shin, Department of Architecture, Architectural Engineering, Hanyang University, Seoul, Korea.

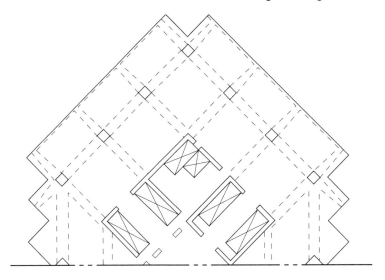

Fig. 4.11 Typical floor plan of 70-storey concrete frame–shear-wall building

spectively. Figure 4.16 shows the differential shortenings at the various floor levels between pairs of columns considered.

The analyses show significant differential shortenings between columns C1 and C3. Of course, for compensation purposes, as mentioned repeatedly, only post-slab-installation differential shortenings are of importance in all-concrete buildings. These are not shown separately for this particular example.

4.10   Sensitivity of Movements Relative to Material Characteristics and Other Factors

As is apparent from the preceding sections, the following variables have an effect on the total shortening of a column.

1. Material characteristics (initial and ultimate values and their time-evolution curves):
    (a) modulus of elasticity;
    (b) shrinkage;
    (c) specific creep.
2. Design parameters:
    (a) cross-sectional area;
    (b) volume-to-surface ratio;
    (c) percentage of reinforcement.
3. Loading parameters:
    (a) progress of construction;
    (b) progress of occupancy;
    (c) environmental conditions (temperature and humidity).

Table 4.2 Properties of exterior column and interior shear wall in 70-storey concrete building (speed of construction: eight days per floor; storey height: 5.5 m bottom storey; 4 m, all other storeys)

Exterior composite column

Floor levels	Concrete strength (MPa)	Col. size (mm × mm)	Gross area (× 10⁶ mm²)	Steel area (percentage gross area)	Floor load* (kN)	Volume-to-surface ratio (mm)	Ultimate shrinkage, (× 10⁻⁶)	Specific creep (× 10⁻⁶ per kPa)
1–10	55	1830 × 1830	3.34	2.47	1132.9	457.2		0.025†
11–20	55	1830 × 1830	3.34	1.23	1132.9	457.2		0.040‡
21–30	55	1525 × 1525	2.32	2.22	1132.9	381.0		
31–40	48	1525 × 1525	2.32	1.78	1132.9	381.0	500†	0.029†
							800‡	0.046‡
41–50	41	1525 × 1525	2.32	1.33	1132.9	381.0		
51–62	41	1120 × 1120	1.25	1.65	1132.9	279.4		
63–67	41	815 × 815	0.66	1.22	1132.9	203.2		0.036†
68–70	41	815 × 815	0.66	1.22	875.4	203.2		0.051‡

Interior shear wall

Floor levels	Concrete strength (MPa)	Wall thickness (mm)	Gross area (× 10⁶ mm²)	Steel area (× 10³ mm²)	Floor load* (kN)	Volume-to-surface ratio (mm)	Ultimate shrinkage (× 10⁻⁶)	Specific creep (× 10⁻⁶ mm per kPa)
1–14	41	610	6.41	70.45	1092.4	288.0		0.036†
15–18	41	508	5.34	57.35	1030.2	242.3		0.051‡
19–30	41	508	5.34	57.35	1061.8	242.3	500†	
31–40	41	406	4.27	45.16	990.1	195.6	800‡	
41–46	31	305	4.27	45.16	990.1	195.6		0.051†
47–72	31	305	3.21	27.10	927.9	148.1		0.065‡

* Dead load plus 0.00048 N mm⁻² live load.
† Low value.
‡ High value.

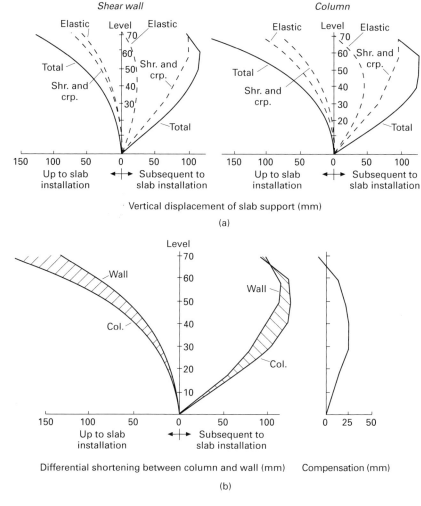

Fig. 4.12 Changes in the lengths of columns and shear walls in 70-storey concrete building, assuming high shrinkage and creep coefficients

Only rarely are the values of ultimate shrinkage and specific creep known at the time of preliminary design when the structural system for a building is determined and when preliminary information on differential shortening is needed to decide on the feasibility of a particular structural configuration.

While the designer has a degree of control over his design assumptions and modulus of elasticity seems reasonably well defined as a function of the specified strength of the concrete, it is necessary to find out during the preliminary design stage how sensitive the potential column length changes may be to variations in

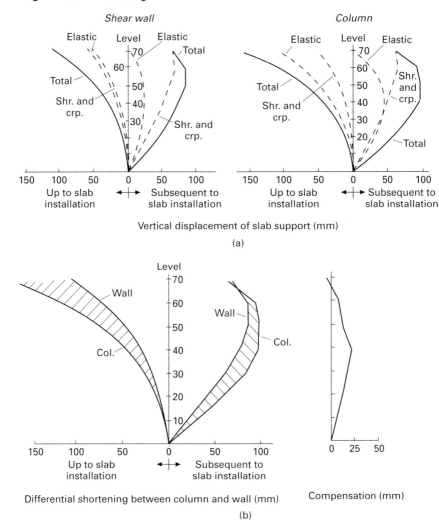

Fig. 4.13 Changes in the lengths of columns and shear walls in 70-storey concrete building, assuming low shrinkage and creep coefficients

the values of ultimate shrinkage and specific creep. Variations in the speed of construction may also have a pronounced effect on the amounts of creep and shrinkage that will occur after a slab is in place.

Sensitivity studies were carried out for the previously described 80-storey composite structure and 70-storey frame–shear-wall interactive structure to determine the effects of low and high values of ultimate shrinkage and specific creep. The results for the 80-storey composite structure are shown in Figs 4.9 and 4.10 in the form of differential length changes between a composite and a steel column. As is seen, the computed tilt of the slabs from their original position in

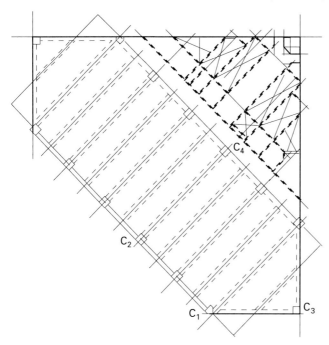

Fig. 4.14 Typical floor plan of 36-storey concrete frame–shear-wall building

the upper storeys of the composite building ranges from 66 to 109 mm for the low and the high shrinkage and creep, respectively, if no precautions are taken in the detailing and erection of the steel columns.

The results for the 70-storey concrete building in the form of differential length changes between a column and a shear wall are presented in Figs 4.12 and 4.13. As is seen, the maximum differential shortenings (at about the 40th floor level) are 22 mm and 27 mm for the low and the high values of shrinkage and creep, respectively.

The results for the two structural systems indicate a high sensitivity of the differential column shortening to values of shrinkage and creep in composite construction and a less than significant sensitivity in all-concrete structures. The reason is that while in the composite building only one of the two differentially shortening elements is subject to shrinkage and creep (the steel column shortens only elastically), both differentially shortening elements in the concrete building shrink and creep.

4.11   Restraining Effect of the Slab System

The differentially shortening supports that have been discussed cause deflections of the supported slab system. In reinforced concrete structures, the deflecting slabs respond to settling supports with resistant shears acting back on the supports, thus decreasing the unrestrained differential shortening. This decrease in

Table 4.3 Properties of column C2 in a 36-storey frame–shear-wall building

Floor levels	Concrete strength (MPa)	Gross col. area (× 10⁶ mm²)	Steel area (× 10³ mm²)	Floor load (kN)	Subsequent floor load (kN)	Volume-to-surface ratio (mm)	Ultimate shrinkage (× 10⁻⁶)	Specific creep (× 10⁻⁶ per kPa)
1–1	49.23	1.96	38.26	812.0	203.0	350.5		
2–2				1118.1	279.5			
3–3	49.23	1.69	38.26	948.4	237.1	325.1		
4–4				1118.1	279.5			
5–5	49.23	1.44	38.26	854.3	213.6	299.7		
6–6				812.1	203.0			
7–7	49.23	1.44	26.77	854.3	213.6	299.7		
8–8				948.4	237.1			
9–10				675.7	168.9			
11–12	49.23	1.44	26.77	675.7	168.9	274.3	650	0.036
13–16				675.7	168.9			
17–22	39.37	1.44	26.77	634.6	158.7	274.3		
23–28	39.37	1.00	15.29	634.6	158.7	248.9		
29–29		0.81	11.48	695.4	173.8	248.9		
30–30				308.0	77.0			
31–31	29.58			675.7	168.9			
32–32				812.1	203.0			
33–33				675.7	168.9			
34–34	29.58	0.49	11.48	484.1	120.1	175.3		
35–35				0.00	0.00			

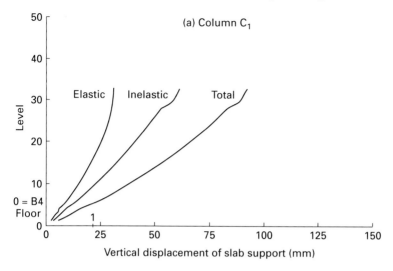

Fig. 4.15a Column length changes in 36-storey concrete frame–shear-wall building

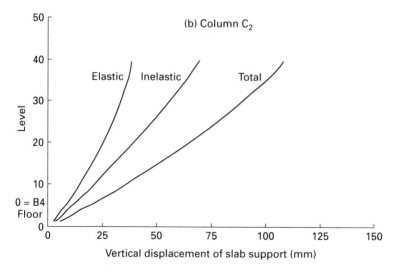

Fig. 4.15b Column length changes in 36-storey concrete frame–shear-wall buildings

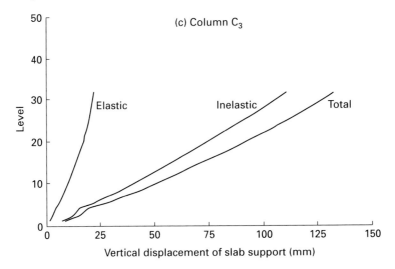

Fig. 4.15c Column length changes in 36-storey concrete frame–shear-wall buildings

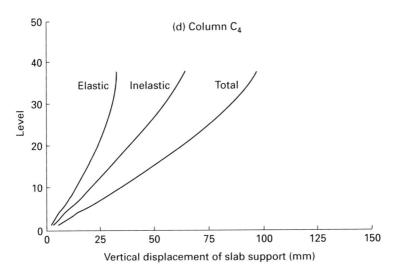

Fig. 4.15d Column length change in 36-storey concrete frame–shear-wall buildings

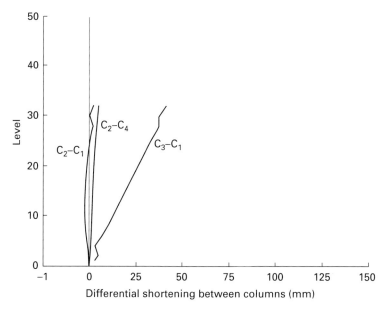

Fig. 4.16 Differential column length changes in 36-storey concrete frame–shear-wall building

shortening is the result of the resistance of the frame and depends on the out-of-plane stiffness of the slab system and the axial stiffness of the columns.

The moments in the slabs due to the differential settlement of supports cause a redistribution of loads between supports, which in turn creates new modified stress levels for creep. A refined consideration of the load redistribution at each stage of construction (after each slab has created, in effect, a new frame) is rarely warranted even in structures with rigid slabs. A relatively simple analysis to determine moments in the slabs due to differential settlement of supports is considered sufficient in view of the existing uncertainties in material properties and stiffness assumptions.

The analysis as given in Ref. 4.5 starts with the known differential settlements of supports (unrestrained by frame action) at each level. A simplification is introduced by substituting a 10-storey one-bay reference frame for the real structure. Reference 4.5 gives graphs to determine the equivalent stiffnesses for the one-bay frame and the resulting residual shortening from which the moments in the slabs and the columns can be determined. As can be seen in the graphs, the residual shortening (as a percentage of unrestrained shortening) depends upon the relative slab-to-column stiffness ratio and upon the number of storeys. For flat-plate-type slabs there is an insignificant restraint resulting in a high residual shortening. With increasing slab stiffness and number of floors, there is less residual shortening.

4.11.1   Effect of Creep on Slab Moments Caused by Differential Settlement of Supports

A restrained member, as shown in Fig. 4.17, subjected to an instantaneous differential settlement of supports Δ, will respond with restraint moments $\pm M$. Creep of the concrete will cause relaxation of moments with time as shown qualitatively by curve A. The change of moment depends upon the creep properties of the member, the change in the effective stiffness of the member caused by progressive cracking, if any, and the increase of the modulus of elasticity of concrete with time.

If the same settlement Δ is applied over a period T, the induced moments will change with time as shown by curve B. The moments reach a maximum at time T and then continue to decrease.

The elastic and inelastic differential shortening of supports of a slab in a multistorey building does not occur instantaneously. The elastic shortening takes place over the period it takes to construct the structure above the slab under consideration. The creep and shrinkage shortening continues for years at a progressively decreasing rate.

The relationship between the magnitude of internal forces caused by settlements and the length of time it takes to apply the settlement was studied experimentally and analytically by Ghali and others.[4.23] The time during which the settlement Δ was applied was varied up to five years. The study shows that for settlements applied during a period of more than 30 days, little change occurs in the maximum value of the parasitic forces, as shown in Fig. 4.18 taken from Ref. 4.23. Based on this study, it seems reasonable to combine the effects on the frame of the elastic and inelastic shortenings of supports.

For practical design of buildings, the author suggests that the maximum value of the differential settlement moments be assumed at 50 per cent of the moments that would occur without relaxation due to creep. These moments should then be used with appropriate load factors in combination with the effects of other loads. The 50 per cent reduction accounts only for creep relaxation during the period of settlement. Beyond this time a further creeping out of settlement moments takes place.

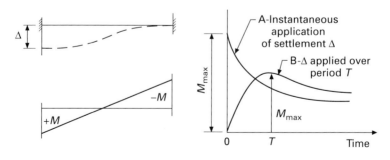

Fig. 4.17 Parasitic moments due to settlement of supports[4.5]

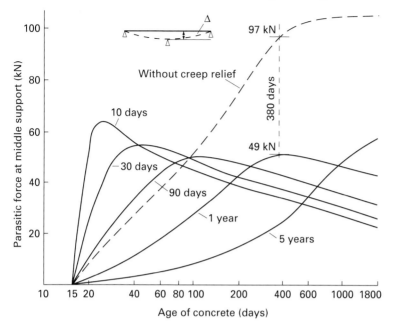

Fig. 4.18 Changes in parasitic forces due to the same settlement occurring over different periods of time[4.23]

4.11.2   Load Transfer Between Adjacent Differentially Shortening Elements

In a frame with differentially settling supports, the support that settles less will receive additional load from the support that settles more. The transferred load is

$$V_i = \frac{M_{i1} - M_{i2}}{L_i}$$ [4.21]

where M_{i1} and M_{i2} are the settlement moments (reduced due to creep relaxation) at the two ends of the horizontal element and L_i is the span.

The load transferred over the entire height of the structure is a summation of loads transferred on all floors. The load transfer is cumulative starting from the top of the structure and progressing down to its base.

4.11.3   Stress

The stresses due to differential shortening should be treated as equivalent dead-load stresses, with appropriate load factors, before combining them with other loading conditions. When choosing a load factor, it should be kept in mind that the design shortening moments occur only for a short time during the life of the structure and they continue to creep after that.

4.12   Relation to Other Permanent and Transient Movements

In addition to the elastic and inelastic length changes due to gravity loads and shrinkage, as discussed in this book, columns and walls in tall structures are subject to length changes due to wind, daily and seasonal temperature variations and foundation settlements.

To establish design criteria specifying limits on distortions that can be accepted at various locations without impairing the strength and the function (serviceability) of the structure, these effects should be combined based on the probabilities of their simultaneous occurrence.

Length changes of columns due to the environmental effects of wind and temperature are transient. Foundation settlements and slab deflections due to gravity loads are permanent and have immediate and long-term components; the latter stabilize with progress of time.

While some of the movements add to the column length changes from elastic stress, creep and shrinkage, other effects (wind) may cause transient elongations that temporarily mitigate the effects of column shortening.

The tilting of slabs due to column length changes caused by shrinkage and gravity loads is permanent; these length changes can and should be compensated for during construction. The permissible tilt of slabs can then be set aside for unavoidable and uncorrectable transient effects of wind and temperature variations.

4.12.1   Wind Movements

Transient wind distortions of a structure basically have no long-term effects. Under wind forces, the columns of a structure are subject to length changes (compression and tension) and to horizontal translation. Only the shortening of the leeward columns is of consequence, since it is in addition to the column shortening caused by gravity loads. In tall and slender buildings this shortening should be considered.

4.12.2   Temperature Movements

Both seasonal and daily temperature fluctuations cause length changes in exterior columns if they are unprotected from the ambient temperatures. The exterior columns elongate when their temperature is higher than the temperature of the interior, protected columns. When these peripheral exposed columns are colder than the interior columns, their thermal shortening is additive to the gravity shortening.

Thermal length changes of columns, like shortening caused by gravity loads, are cumulative. Therefore, cooling of the exposed columns in tall structures (during the winter) combined with column shortenings due to gravity may become a critical design consideration. The daily temperature fluctuations have a less aggravating effect, because the rapid temperature fluctuations do not penetrate sufficiently into the depth of the concrete to cause length changes. Only

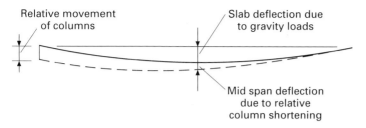

Fig. 4.19 Combined slab deflections due to gravity loads and column shortening

average temperatures over a number of days (depending on the column size) can have an effect on column length changes.

4.12.3   Differential Foundation Movements

Relative foundation settlements that occur between the core and the exterior peripheral columns need to be considered, along with the differential column shortenings due to shrinkage and gravity loads, to determine the combined effects on the slab systems. In most cases a larger settlement occurs at the centre of a building, regardless of whether individual footings or mat foundations are used. On the other hand, the gravity effects in reinforced concrete buildings usually cause larger shortenings of the peripheral columns relative to the core walls. Only a precise estimate of the two types of distortion can determine if they cancel each other and whether their superposition can be omitted. It seems practical to consider each of the two effects separately, as if the other did not exist. This is advisable in view of the uncertainty in realistic prediction of differential foundation settlement. Rarely in the past could the predicted amount of differential foundation settlement be verified by actual field measurements. Further field observations of foundation settlement are needed before they can be considered in combination with other effects.

4.12.4   Vertical Deflections of Slabs Due to Gravity

Figure 4.19 shows the superposition of slab deflections due to gravity loads on slabs and those due to differential shortening of the supporting columns. As is seen, the column shortening increases the tilt of the slab and therefore has an effect on partition distortions.

4.13   Performance Criteria – Limitation on Distortions

Effects of the tilt of slabs on partitions and finishes require limitations on column movements, although it is possible to structurally accommodate a considerable amount of column movement either by reinforcing for the effects of such movement or by having the slabs hinged. Serviceability criteria for slabs are expressed in terms of the ratio of the support settlement to the slab span length (angular distortion), as well as the absolute amount of movement.

From the previous discussion, it is evident that in considering limitations, the column shortenings due to gravity loads and shrinkage should be superimposed with other effects and limitations established on the combined movement. As is seen in Fig. 4.19, the column shortening due to gravity and that due to thermal effects and wind can be additive to vertical load deflections of the slab in their effects on partition distortions and on slab tilt.

Traditional British practice limits angular distortions caused by differential settlements to $L/360$ (L is the span length of member) to protect brickwork and plaster from cracking; $L/180$ is considered tolerable in warehouses and industrial buildings without masonry. Although in modern high-rise buildings the considerations of performance are totally different from that of cracking of masonry and plaster (these materials having almost disappeared from modern buildings), limitations such as those in British practice may still be suitable, since they result in distortion limits customary in the construction industry. Thus, a suggested limitation of $L/240$ to be applied to the combined differential column shortening due to elastic stresses, creep, shrinkage, temperature variations and wind would limit the slab distortion at mid-span to $L/480$ plus the slab deflection due to gravity (see Fig. 4.19).

In keeping with standards used in some leading engineering offices, additional limits on the maximum differential settlement of supports of a slab, applicable regardless of the span length, can be suggested. The most commonly used limit appears to be 20 mm on the differential settlement caused by either temperature or wind or elastic stresses plus creep plus shrinkage and 25 mm on the differential settlement caused by a combination of all the above factors.

4.14   Compensation for Differential Support Shortening

The main purpose in computing anticipated column shortenings is to compensate for the differential length changes during construction so as to ensure that the slabs will be horizontal in their final position. Such compensation for differential length changes is similar in concept to the cambering of slabs, since both processes are intended to offset deformations anticipated in the future.

At the time a slab is installed, each of the two slab supports has already undergone a certain amount of shortening. After slab installation, the supports will settle further, due to subsequent loads as construction progresses and due to shrinkage and creep. In composite and concrete structures where the anticipated shortenings may take place over many years, full compensation can be made during construction, provided the amount is not excessive. Where large differential length changes need to be compensated for, compensation can be made during construction to offset movements expected to take place over several years; subsequent shrinkage and creep will then cause tilt of the slab. Note that during the initial two years about 85 to 90 per cent of all shrinkage and creep distortions will have taken place.

The methods of compensation are different in concrete structures and in each of the two types of composite structures.

4.14.1   Concrete Structures

In concrete structures, compensation is required only for the differential settlements of supports that are expected to occur after slab installation. Each time formwork for a new slab is erected, its position is adjusted to the desired level; the pre-installation settlements of the supports are thus eliminated.

The anticipated differential shortening of the slab supports to be compensated for consists of the following components.

1. Elastic: due to
    (a) loads above (progress of construction);
    (b) loads of finishes below and above;
    (c) live loads below and above.
2. Inelastic: due to shrinkage and creep from the time of slab installation.

Either complete or partial compensation of the differential length changes can be specified for the formwork, depending upon the magnitude of anticipated support settlements.

4.14.2   Composite Structures

Composite structures in which some of the columns are steel and others are concrete or composite have a considerably high potential for differential shortening, as seen previously. Therefore, neglecting to consider vertical distortions in very tall composite structures may cause behavioural problems, either immediately or sometimes years after the building has been in satisfactory operation.

In composite structures consisting of interior steel columns and an exterior reinforced concrete beam–column system with embedded erection columns, the steel columns are usually fabricated to specified storey-high lengths. Traditionally, after a number of floors (say, six to 10) have been erected, the elevations of the column tops are measured and shim plates to compensate for the differential length changes are added, to start all columns of the new storey from the same elevation. Thus, compensation is made for pre-slab-installation shortenings; but post-slab-installation shortenings will cause tilting of the slabs. If an analysis for column shortening is available, it may be possible to upgrade this traditional procedure by modifying the thickness of the shim plates to include the differential shortenings anticipated after slab installation.

To avoid the costly shim-plate procedure and to ensure horizontal slabs in the finished building, column lengths can be detailed to compensate for the anticipated shortening due to all loads. The compensation can be detailed for every column lift; or if the computed column-lift differential is smaller than the fabrication tolerances, then after a number of floors (say, six to 10), a lift with corrected column lengths can be detailed.

As seen in Figs 4.9 and 4.10, the differential shortening of columns in the 80-storey composite structure includes the shortening of the erection columns prior to their embedment into the concrete in addition to the elastic and inelastic shortenings of the concrete in the composite columns. On the right of Figs 4.9(c) and 4.10(b), the suggested modification of the lengths of the steel erection columns is shown. While the erection column at the 10th storey needs to be longer by 27 mm according to Fig. 4.10(b), this differential becomes gradually smaller and is 10 mm at the 60th storey and 4 mm at the 70th storey. The erection column becomes shorter than the interior steel column by 3 mm at the 80th storey.

In composite structures consisting of an interior concrete core and steel peripheral columns and steel slab beams, compensation for column and wall length changes can be made by attaching the support plates for the steel slab beams to the core (into pockets left in the wall), regardless of whether the core is constructed with jump forms proceeding simultaneously with steel erection or is slipformed. In both cases, the before-slab-installation shortening is eliminated if the steel beams are installed horizontally.

In jump-formed core construction, the beam support plates can be embedded in the pockets at the desired location to accommodate anticipated support shortening after the slab has been installed, either by attaching them to the forms or by placing them on mortar beds after the wall is cast.

In slipformed cores, pockets are left to receive the slab beams. Sufficient tolerance needs to be detailed in the size and location of the pockets so that the beams can be placed to accommodate the anticipated shortening of supports.

With the above arrangements, the steel columns do not need compensation and can be manufactured to the specified storey-high length; only the elevation of the support plates for the beam ends resting on the concrete walls needs adjustment to provide for the differential shortening that will occur after the slab is in place.

4.15   Testing of Materials to Acquire Data

Information on shrinkage, creep and modulus of elasticity for various strengths of concrete for use in preliminary analyses of column shortening has been discussed. The information is based on American and European literature.

Should preliminary analyses indicate substantial differential length changes, it may be desirable for a more refined prediction of such length changes to determine the material characteristics of the actual concretes to be used in a laboratory testing programme prior to the start of construction.

Standard 150 mm by 300 mm cylinders made of the actual design mixes should be used to determine the strength, modulus of elasticity, shrinkage and creep coefficients and the coefficient of thermal expansion. All specimens should be moist-cured for seven days and then stored in an atmosphere of 23°C and 50 per cent relative humidity. A modest testing programme would entail a total of 12 test cylinders for each of the concrete strengths to be used in the structure: six cylinders for strength (two each for testing at 28, 90 and 180 days); two cylinders for the shrinkage coefficient; two cylinders for the creep coefficient; one sealed cylinder (in copper foil) for thermal expansion; and one spare cylinder.

The creep specimens should be loaded at 28 days of age to a stress level expected in the structure, but not to exceed 40 per cent of the cylinder strength.

Measurement of shrinkage on the shrinkage specimen should start after seven days of moist-curing. While the measurements can go on for several years, the 90-day interim readings can be used to extrapolate the ultimate values of shrinkage and creep, although the extrapolation will obviously be more accurate for data obtained over longer periods of time. The thermal coefficient should be measured on the sealed specimen at 28 days between temperatures of 4°C and 38°C.

4.16   Verification

Two 152 × 152 × 914 mm lightweight concrete columns reinforced with four no. 5 (16 mm-diameter) deformed 517 MPa bars were tested in the laboratories of the Portland Cement Association under 50 equal weekly increments of load, each equalling 2 per cent of the full load of 311 kN, with the first increment applied at one week of age.[4.24] The load simulated conditions encountered in a 50-storey concrete building. The columns were moist-cured for three days and then sealed in thick copper foil to simulate idealized moisture conditions in large prototype columns. Such sealing also virtually eliminated shrinkage. The measured data from the incrementally loaded columns are shown in Fig. 4.20. Column strains were measured on three faces of each column, over 254 mm gauge lengths at mid-height.

Comparison cylinders, 150 × 300 mm, were cast and continuously moist-cured until time of testing for compressive strength, modulus of elasticity, creep and drying shrinkage characteristics as functions of time. Creep tests were conducted on sealed as well as unsealed specimens. The measured 28-day compressive strength of concrete was 44 MPa and the measured modulus of elasticity was 23 029 MPa. Both values represent average results from four cylinder tests. After 146 weeks of loading, the measured creep strains of sealed and unsealed cylinders loaded at 28 days of age were 250 and 725×10^{-6}, respectively, under 10.34 MPa of stress. The corresponding projected ultimate values (using Eq. 4.10) of sealed and unsealed creep are 289 and 838×10^{-6}, respectively, for

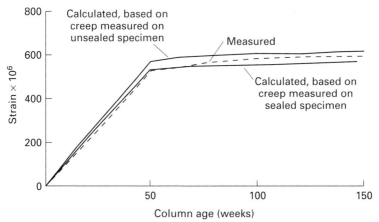

Fig. 4.20 Verification of proposed analytical procedure

10.34 MPa of stress. These translate into sealed and unsealed specific creep values of 0.280×10^{-7} per kPa 0.881×10^{-7} per kPa, respectively.

Elastic plus creep strains for the test columns were predicted using the procedures outlined in this chapter. Measured values of compressive strength and modulus of elasticity were used in the computations. Two sets of computations were made – one using the unsealed specific creep value of 0.881×10^{-7} per kPa with a corresponding volume-to-surface ratio of 38 mm and the other using the sealed specific creep value of 0.280×10^{-7} per kPa with a corresponding large volume-to-surface ratio of 254 mm. Since the test columns were sealed, shrinkage was not considered in the computations. The analytical results are compared with the test data in Fig. 4.20. Excellent agreement is observed, irrespective of whether unsealed or sealed creep values are used in computations. It is obviously much simpler to test unsealed, rather than sealed, specimens. Testing of sealed specimens appears to be unnecessary under most circumstances. Realistic prediction of creep deformations seems to be possible on the basis of creep measured on unsealed specimens.

Figure 4.20 provides verification of the proposed column strain prediction procedure including various components of the creep model. A direct verification of the shrinkage model, however, was not obtained because the test columns were sealed.

Reference 4.6 provides comparisons between measured and predicted column strains for all six instrumented levels of a particular column and for all five instrumented columns on a particular level of Water Tower Place, a 76-storey reinforced concrete building located in Chicago, Illinois (until 1989, the tallest concrete building in the world). Encouraging agreement between measurement and prediction was observed. Reasonable agreement between actual measured values and analytically predicted values of strains were also observed for two isolated columns and two shear walls on the second floor of 3150 Lake Shore Drive, Chicago, a 34-storey reinforced concrete building. Instrumentation has been installed on a number of newer buildings like One Magnificent Mile in Chicago, Southeast Financial Center in Miami and 311 South Wacker in Chicago. In each case, column shortening predictions have been made using the methodology presented in this chapter. Comparisons between measured and computed deformations, which are understood to have been very close at least in the case of 311 South Wacker, are not available to the author at the time of writing. Instrumentation is now being installed on one of the twin towers of Kuala Lumpur City Centre, the soon-to-be tallest concrete building in the world. Analysis of column shortening will again be made using the methodology presented herein. Comparisons between measured and predicted column deformations are expected to be available in the future.

4.17   Summary and Conclusions

In high-rise buildings, the total elastic and inelastic shortening of columns and walls due to gravity loads and shrinkage may be as high as 25 mm for every 24 m of height. The possibly large absolute amount of cumulative column shortening over the height of the structure in ultra-high-rise buildings is of consequence in its effects on the cladding, finishes, partitions etc. These effects can be contained by providing details at every floor level that would allow the vertical structural members to deform without stressing the non-structural elements. The differential shortening between adjacent columns may cause distortion of slabs, leading to impaired serviceability. The tilting of slabs due to length changes of columns caused by gravity loads and shrinkage are permanent; they should be compensated for during construction. The permissible tilt of slabs can then be set aside for the unavoidable and uncorrectable transient effects of wind and temperature.

An analytical procedure has been developed to predict the anticipated elastic and inelastic shortenings that will occur in the columns and the walls both before a slab is installed and after its installation. This procedure has been verified and found to be in reasonable agreement with a number of field measurements of column shortening in tall structures that extended over periods of up to 19 years.[4.6]

It should be noted that a differential between two summations, each of which consists of many components, is very sensitive to the accuracy of the individual components. A small change in an individual component may substantially alter the differential between the two summations.

Design examples of column shortening have been presented for the following structural systems used for ultra-high-rise buildings:

- reinforced concrete frame–shear-wall system;
- composite system, with peripheral reinforced concrete beam–column frames.

The differential shortening between the columns and walls of the 70-storey reinforced concrete structure studied was 22 and 27 mm, for assumed low and high values of shrinkage and creep, respectively. For the 80-storey composite structure investigated, the predicted differential shortenings between the reinforced concrete and the steel columns were 66 and 109 mm, for assumed low and high values of shrinkage and creep, respectively. Obviously, such differential length changes should be compensated for during construction and the fabricated column detailed accordingly.

The column shortenings predicted by the analytical procedure proposed here are computed column shortenings in the absence of any restraint caused by frame action. In a general case, these computed unrestrained column shortenings are to be input into a general structural analysis program which in turn will compute the actual restrained (residual) column shortenings.

Limitations on slab distortions caused by column shortening due to gravity loads and shrinkage effects are recommended. For several different structural systems, different construction methods to compensate for differential shortenings are suggested.

Compensation for differential column shortening in reinforced concrete structures is similar to cambering and is relatively simple: the formwork along one edge is raised by the specified amount.

In composite structures with a reinforced concrete peripheral beam–column bracing system, the steel columns are detailed to compensate for all length changes.

In composite structures braced with a reinforced concrete core, differential length changes are compensated for by adjusting the embedment levels of the support plates of the steel beams in the core; the steel columns can be fabricated to the specified storey heights.

The apparent greater sensitivity of composite structures to differential shortening makes it desirable to consider such movements and their effects on the slab during design and to plan steps to be taken during construction to limit slab tilt to 20 mm or less.

Acknowledgement

The author wishes to acknowledge valuable collaboration with Mark Fintel during the development of the methodology reported on in this chapter.

References

4.1   Fintel M, Kahn F R 1965 Effects of column exposure in tall structures – temperature variations and their effects. *ACI Journal, Proceedings* **62**(12): 1533–1556

4.2   Kahn F R, Fintel M 1966 Effects of column exposure in tall structures – analysis for length changes of exposed columns. *ACI Journal, Proceedings* **63**(8): 843–864

4.3   Kahn F R, Fintel M 1968 Effects of column exposure in tall structures – design considerations and field observations of buildings. *ACI Journal, Proceedings* **65**(2): 99–110

4.4   Fintel M, Khan F R 1969 Effects of column creep and shrinkage in tall structures – prediction of inelastic column shortening. *ACI Journal, Proceedings* **66**(12): 957–967

4.5   Fintel M, Khan F R 1971 Effects of column creep and shrinkage in tall structures – analysis for differential shortening of columns and field observation of structures. *Symposium on Designing for Effects of Creep, Shrinkage, and Temperature in Concrete Structures* Publication SP-27, American Concrete Institute, Detroit, MI, pp 159–185

4.6   Fintel M, Iyengar S H, Ghosh S K 1986 *Column shortening in tall structures – prediction and compensation* Publication EB108D, Portland Cement Association, Skokie, IL

4.7   American Concrete Institute 1992 *Building Code Requirements for Reinforced Concrete* ACI 318-89 (Revised 1992)

4.8   ACI Committee 209 1971 Prediction of creep, shrinkage, and temperature effects in concrete structures. *Designing for Effects of Creep, Shrinkage, and Temperature in Concrete Structures* Publication SP-27, American Concrete Institute, Detroit, MI, 51–93

4.9   Martinez S, Nilson A H, Slate F O 1982 *Spirally Reinforced High-Strength Concrete Columns* Research Report No. 82-10, Department of Structural Engineering, Cornell University, Ithaca, NY

4.10   ACI Committee 363 1992 *State-of-the-Art Report on High-Strength Concrete* Publication 363R-92, American Concrete Institute, Detroit, MI

4.11   RILEM Commission 42-CEA 1981 Properties of set concrete at early ages: state-of-the-art report. *Materials and Structures* **14**(84): 399–450

4.12   Byfors J 1980 *Plain Concrete at Early Ages* Swedish Cement and Concrete Research Institute, Fo. 3, No. 80, Stockholm, Sweden

4.13   Russell H G, Corley W G 1977 *Time-Dependent Behaviour of Columns in Water Tower Place* Publication RD052B, Portland Cement Association, Skokie, IL

4.14   Troxell G E, Davis H E, Kelly J W 1968 *Composition and Properties of Concrete* 2nd edn, McGraw-Hill Book Co., New York

4.15   Hansen T C, Mattock A H 1966 Influence of size and shape of member on the shrinkage and creep of concrete. *ACI Journal, Proceedings* **63**(2): 267–289

4.16   American Concrete Institute, undated *Recommendations for an International Code of Practice for Reinforced Concrete* American Concrete Institute, Detroit, MI and Cement and Concrete Association, London

4.17   CEB 1972 *Manual: Structural Effects of Time-Dependent Behaviour of Concrete* Bulletin d'Information, No. 80, Comité Europeen du Béton (CEB), Paris

4.18   Hickey K B 1968 *Creep of Concrete Predicted from Elastic Modulus Tests* Report No. C-1242, United States Department of the Interior, Bureau of Reclamation, Denver, CO

4.19   Neville A M 1981 *Properties of Concrete* 3rd edn, Pitman, London

4.20   Pfeifer D W, Magura D D, Russell H G, Corley W G 1971 Time-dependent deformations in a 70-storey structure. *Symposium on Designing for Effects of Creep, Shrinkage, and Temperature in Concrete Structures* Publication SP-27, American Concrete Institute, Detroit, MI, pp 159–185

4.21   Troxell G E, Raphael J M, Davis R E 1958 Long-time creep and shrinkage tests of plain and reinforced concrete. *Proceedings American Society for Testing and Materials* **58**: 1101–1120

4.22   Dischinger F 1937 *Der Bauingenieur* Berlin, October 1937; also December 1939

4.23   Ghali A, Dilger W, Neville A M 1969 Time-dependent forces induced by settlement of supports in continuous reinforced concrete beams. *ACI Journal, Proceedings* **66**(11): 907–915

4.24   Pfeifer D W, Hognestad E 1971 Incremental loading of reinforced lightweight concrete columns. *Lightweight Concrete* Publication SP-29, American Concrete Institute, Detroit, MI, pp 35–45

5 Rational Methods for Detailing and Design: Strut-and-Tie Modelling

Karl-Heinz Reineck

Extensive research in the last decade, especially into shear problems, has provided a new impetus to the use of rational approaches for the design of structural concrete through the concepts of strut-and-tie modelling. The fundamentals of this design approach are explained in this chapter, and the method is illustrated with some typical design problems in B-regions and D-regions.

5.1 Introduction

In the beginnings of reinforced concrete in the late 19th century, engineers like Hennebique and Maillart recognized the advantages of the new material and dared to build impressive structures with little guidance from codes. They based their designs on the simple and basic idea of using the concrete to carry compression, and the steel reinforcement to carry any tensile forces. This idea was further developed by Ritter[5.1] and Mörsch[5.2,5.3] into the truss analogy for the analysis and design of beams subjected to bending and shear.

Today, the simple basic idea tends to be obscured by the bulk of rules in codes and handbooks, which are often not properly understood by designers. Admittedly there were inadequacies in the models available to treat shear and punching in members without transverse reinforcement, and to deal with bond and anchorage. Designers have therefore become accustomed to empirical rules expressed in terms of shear stresses and bond stresses. We recognize nowadays that not all tensile forces can be taken by the reinforcement, and that the tensile strength of concrete is of fundamental importance in many design situations.

Structural damage such as that reported by Leonhardt[5.4,5.5] and Podolny,[5.6] and the recent spectacular loss of an entire offshore platform,[5.7–5.9] have shown that it is unsatisfactory to rely on empirical approaches and rules of thumb for the design of important regions of a structure such as those with static or geometric discontinuities. In such regions the standard design rules for flexure and shear do not apply. To make such unsatisfactory and risky empirical procedures unnecessary, Schlaich[5.10–5.12] as well as Thürlimann,[5.13–5.15] Marti,[5.16] Stone and Breen,[5.17] and Collins and Mitchell[5.18–5.20] have proposed a generalization of the truss analogy for the design of structural concrete, with the consistent application of 'strut-and-tie models' to all regions of a structure. The strut-and-tie approach had previously been used for design problems such as deep beams and corbels[5.21] and even for beam–column connections under severe earthquake loads.[5.22,5.23]

At the 1991 IABSE Colloquium on Structural Concrete in Stuttgart, further emphasis was given to the need for consistent and transparent models for designing and detailing concrete structures.[5.24,5.25] The importance of considering overall behaviour and the flow of forces in structures was also emphasized, with a further call for generalized approaches which cover the entire spectrum of structural concrete.[5.26,5.27]

In following up the aims of the 1991 Colloquium, the present chapter presents current information on rational design procedures and on strut-and-tie modelling. The intention is to demonstrate that this approach can cover all design problems, including currently disputed problem areas related to shear design and concrete tensile strength. The design of some typical discontinuity regions is demonstrated with models and examples.

5.2   General Procedure for Strut-and-Tie Modelling

5.2.1   B-Regions and D-Regions

Those regions of a structure in which Bernoulli's hypothesis of plane distribution of strain can be applied have been referred to by Schlaich[5.11] as *B-regions*, where the B may be taken to stand for *B*eam or for *B*ernoulli. The fact that these regions can be designed using simple truss models tends to be hidden by code rules for cross-section design which deal separately with the action effects (bending, axial force, shear etc.) and their various combinations.

The standard methods are not applicable to specific regions where the strain distribution is significantly non-linear, for example at concentrated loads, in corners of frames, at connections and bends, in corbels, recesses and openings, and where prestressing tendons are anchored. Examples are shown in Fig. 5.1. Such regions are called *D-regions*, whereby the D may be considered to stand for *D*iscontinuity, *D*isturbance or *D*etail. The D-regions are static or geometrical discontinuities, or combinations of the two, and are of limited extent according to Saint-Vénant's Principle.[5.11, 5.28] The division of a structure into B- and D-regions (Fig. 5.2) can contribute considerably to the understanding of the flow of the forces, even before any structural analysis is undertaken.

5.2.2   General Design Procedure

Before a strut-and-tie model (STM) can be developed for the B- and D-regions of a structure, a global structural analysis must be carried out in order to determine the support forces and maximum moments which are used to dimension each member and region. In structural design, which comprises analysis, dimensioning and detailing, one must comply with the requirements of equilibrium, compatibility and the strength and deformation properties of the materials. However, these requirements cannot always be fulfilled rigorously, and simplifications are often introduced, especially in regard to material behaviour. During the design process, various analysis and design models may be used with differing degrees of sophistication, according to the type of structure, the required information and the required accuracy.[5.27]

In practice, a linear-elastic analysis is normally used, even though this is really only valid for structural concrete which is uncracked, and provided the com-

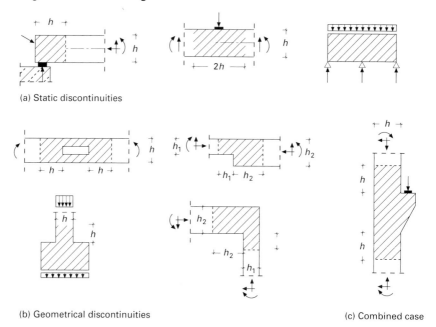

(a) Static discontinuities

(b) Geometrical discontinuities   (c) Combined case

Fig. 5.1 Typical D-regions with non-linear strain distributions due to static and geometrical discontinuities[5.11]

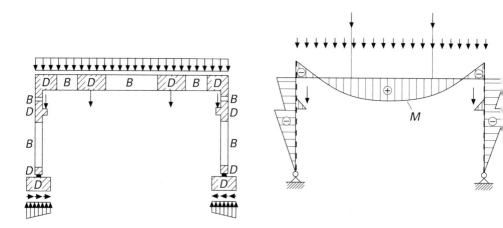

Fig. 5.2 Frame with its B- and D-regions compared with the static system with the bending moments from a beam analysis[5.11]

pressive stresses are not excessive (e.g. $\leq 0.4f_c$). Non-linear analysis is rarely used in practice because of the large amount of labour required; it is only applied to unusual structures and in research.[5.27] In any case, such an analysis can only be carried out when the dimensions and details for the concrete and the steel have already been determined.

Although the theory of plasticity, as applied by Thürlimann *et al.*[5.14,5.15] and Nielsen *et al.*[5.29] grossly simplifies the behaviour of structural concrete, it leads to a simple and clear theory which is applicable to different materials. Of special importance is the static method, which is often used in strut-and-tie modelling. However, there are limitations to the application of the theory of plasticity to structural concrete. Examples are given below in Sections 5.3.3 and 5.5. In applying the theory of plasticity it is always necessary to check that ductile behaviour can in fact be guaranteed.

5.2.3   Modelling Procedure

In modelling B-regions and D-regions, a system of struts and ties has to be found which is in equilibrium with the given loads and represents the behaviour of the structure. The basic elements of a strut-and-tie model (STM) are shown in the examples in Fig. 5.3, where:

- a strut represents the resultant of a parallel or fan-shaped compressive stress field;
- a tie represents the resultant of forces in the reinforcement or in a concrete tensile zone;
- a node is a confined region of concrete in which the compressive strut forces are equilibrated by anchored reinforcement or other struts.

The static system of the STM may also contain nodes within the D-regions, but these are diffused through relatively large areas as biaxial stress fields and are not therefore treated as nodes here.

When the model has been constructed, it is necessary to evaluate the tie forces, the tensile and compressive stresses in the concrete, and the stresses in the nodes and anchorages. In determining the concrete stresses, it is often possible and advantageous to model the stress fields directly in the manner shown by Muttoni *et al.*[5.30]

The indispensable requirement for any model is to satisfy the requirements of equilibrium, and the material strength limits. This implies that the structure is designed according to the lower bound theorem of the theory of plasticity (static method). As a simple means of also allowing for compatibility, Schlaich *et al.*[5.11] have recommended the use of elastic analysis as a basis for defining the strut-and-tie model. If the results of a linear-elastic analysis are not available, 'standard models' can be created from comparable cases, or alternatively the STM can be developed using the so-called 'load-path method'. The basic idea of the load-path method is to balance the resultants of the stress diagram acting at one border of a B-region with their counterparts at the other border, as well as any loads applied within the D-region. Load paths are then used to connect the forces from the

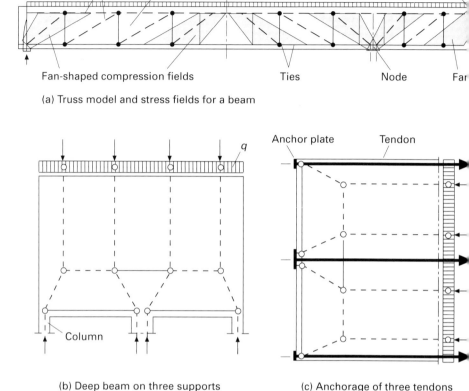

(a) Truss model and stress fields for a beam

(b) Deep beam on three supports

(c) Anchorage of three tendons in a slab bridge

Note: the same strut-and-tie model appears in (b) and (c) for two different D-regions

Fig. 5.3 Typical examples for strut-and-tie models and stress fields

opposite borders, with diversions made as necessary at nodes by means of transverse ties and struts. The procedure has been illustrated by Schlaich *et al.*[5.11] A further possibility for modelling has recently been suggested by Rückert,[5.31] who implemented strut-and-tie modelling in a CAD system linked to analysis and design programs and run under a uniform interface.

A semi-automatic interactive method is available in which the STM is developed from linear-elastic stress trajectory fields. The resulting model is kinematic, as in many cases of strut-and-tie models, so that equilibrium is only possible for a specific geometry. The difficulty of analysing such models has been solved by using a geometrically non-linear program, which uses initial stiffnesses. This powerful tool exploits the numerical and graphical capabilities of the computer and allows integration of the strut-and-tie method into computer-orientated design procedures.

Finally it should be noted that strut-and-tie modelling makes obvious that dimensioning and detailing are closely related. Many of the 'detailing rules' which are dealt with on an *ad hoc* basis in codes, become an integral part of modelling and dimensioning with the use of the STM. This follows from the modelling requirement that the tie forces must coincide with the direction and location of the reinforcement. An iterative procedure can thus be used to adjust the model to the reinforcement layout and to the dimensions of the nodes and anchorages.

5.3   Strength of the Components of the Strut-and-Tie Models

5.3.1   Steel Ties

For the ultimate limit state it is assumed that the tension steel yields, so that the capacity of the tie is:

- for reinforcing steel: $F_{syd} = A_s f_{yd}$;
- for prestressing steel: $F_{pyd} = A_p f_{pyd}$.

The latter case is only valid if there is sufficient prestrain to ensure that the additional stresses due to loading bring the tendon to the yield stress, f_{pyd}. If the prestress is applied as external load in the analysis, only the reserve capacity between the prestressing force and the yield force can be utilized.

5.3.2   Struts

A compressive strut may represent a parallelstress field (Fig. 5.4a), or a fan-shaped stress field (Fig. 5.4b). These have already been shown in beam applications in Fig 5.3(a). The basis for the design strength of a strut is the uniaxial design strength of the concrete in compression, f_{1cd}, as defined in relevant codes and standards such as Eurocode EC 2, Part 1[5.32] and CEB-FIP MC 90:[5.33]

$$f_{1cd} = 0.85 f_{cd} = 0.85(f_{ck}/\gamma_c)$$

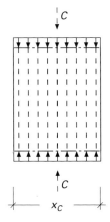

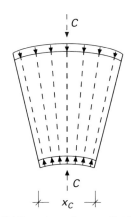

(a) Parallel or prismatic stress field   (b) Fan-shaped stress field

Fig. 5.4 Typical compression stress fields for struts

Here f_{ck} is the characteristic cylinder strength, and γ_c is the partial safety factor for concrete (usually taken as 1.5). The effective concrete strength in a strut is smaller than this value for various reasons, and is expressed using two reduction factors, v_1 and v_2:

$$f_{cd,eff} = v_1 f_{lcd} \quad \text{or} \quad f_{cd,eff} = v_2 f_{lcd}$$

The reduction factor v_1 applies to a prismatic strut in undisturbed and un-cracked regions. It allows for the fact that $f_{cd,eff}$ normally applies to a rectangular stress block rather than to the real stress distribution. In the case of a linear strain gradient, as in the compression chord in a beam, there are alternative ways for calculating the resultant force by means of a stress block. If an average or effective stress is taken over the full depth of the compression zone,[5.33] then the appropriate value is

$$v_1 = 1 - \frac{f_{ck}}{250}$$

It should be noted that for high-strength concrete there is actually not a reduction in the strength itself, but in the ultimate strain, and this reflects in the total force integrated over the compression zone. If the stress distribution in the strut is determined from the compatibility of strains by using realistic stress–strain laws, then of course the factor is $v_1 = 1.0$.

The factor v_2 applies to parallel or fan-shaped stress fields and its value depends on various influencing effects. A first group of effects, which cannot be separated, is summarized in Fig. 5.5, and in the first instance takes account of the transverse tension and geometric influences due to disturbances. Various researchers found that the maximum reduction is from 15 to 20 per cent.[5.34–5.36]

Lower values for the strength of struts are only justified if the compression field contains cracks which cross the struts (Fig. 5.6). The reduction then depends on the width of the cracks because the force transfer across the cracks requires the frictional resistance of the rough crack surfaces, and this reduces with increasing crack width. The crack width itself depends on the transverse strain.[5.18] However, this is not a complete explanation, because the spacing of the cracks also plays an important role.[5.37] For very large crack spacing or large crack width, the reduction can be large and this can become relevant for tension flanges in torsion.

In summarizing all these influencing parameters, the following simplified values are proposed:

- $v_2 = 0.80$ for struts crossed by ties and with cracks parallel to the direction of the strut;
- $v_2 = 0.60$ for struts transferring compression over cracks with normal crack widths;
- $v_2 = 0.40$ for struts transferring compression over large cracks (e.g. members with axial tension, such as tension flanges of box girders).

Normally such refined considerations are not relevant, and the value $v_2 = 0.60$ may be used in designing D-regions because compression is rarely decisive in well-designed structures.

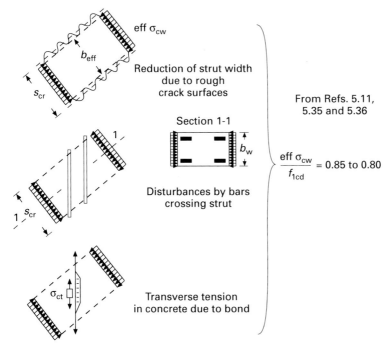

Fig. 5.5 Causes for strength reduction in a strut parallel to cracks[5.37]

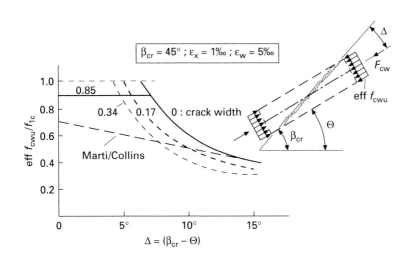

Fig. 5.6 Strength of struts crossed by cracks[5.37]

5.3.3   Concrete 'Tensile Ties'

The ultimate capacity of a member should not rely on the tensile resistance of concrete, but instead on reinforcement. However, this is not completely true, because the concrete tensile strength actually has to be utilized in members without transverse reinforcement and also for bond and anchorage. Furthermore, it is advantageous to take account of any uncracked regions and the tension stiffening effect of concrete in cracked regions of a member in the global analysis which is used as the basis for the STM. Crack formation often plays an important role at the ultimate limit state, and it may be assessed in a simple way. The characteristic value of the axial tensile strength is not compared with the maximum stress in a section, but with the average tensile stress over a zone of three times the maximum aggregate size, but not more than 50 mm. This can replace the empirical formulae otherwise used.

In modelling uncracked regions, biaxial tension–compression fields have to be considered. The parallel biaxial stress field in Fig. 5.7 may represent uncracked regions, or alternatively the behaviour of the concrete between cracks, as in the case of webs with transverse reinforcement or slabs without transverse reinforcement, as discussed later in Sections 5.4 and 5.5.

The other biaxial tension–compression stress field in Fig. 5.8 is bottle-shaped. It may be used for modelling uncracked D-regions. Its capacity depends very much on the ratio of the support width a to the total width b. The lowest value for the applied pressure, $p_a = 0.60 f_{ct}$, is reached with a value of about $a/b = 0.5$, and is governed by the transverse tensile stress f_{ct}. This is a safe value for the cracking load of the bottle in Fig. 5.8.

5.3.4   Nodes and Anchorages

The nodes must be dimensioned and detailed so that all forces are balanced and the ties are anchored safely. The concrete is biaxially or triaxially stressed. The

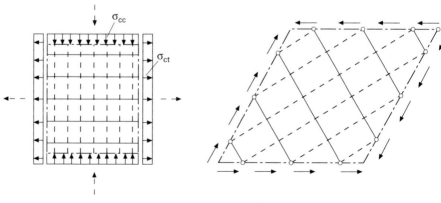

(a) Rectangular plate element   (b) Skew plate element for webs

Fig. 5.7 Parallel biaxial tension–compression fields in the concrete

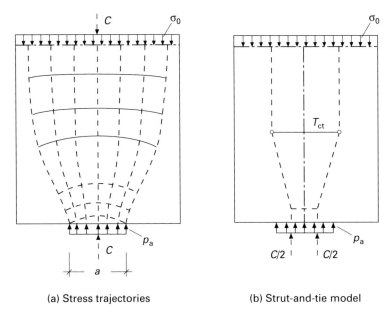

(a) Stress trajectories   (b) Strut-and-tie model

Fig.5.8 Bottle-shaped tension–compression field in the concrete[5.12]

stress may be purely compression, as in nodes connecting struts, or compression and tension if bonded reinforcement is anchored. The adequacy of the nodes may be verified using the following steps.

1. Check the adequacy of the anchorage of any ties in the node. The anchorage length is defined by the points where the deviations of the compressive field, as caused by the reinforcement, start and end.
2. Ensure that the maximum compressive stress does not exceed the effective strength. In nodes connecting only compressive struts the biaxial or triaxial compressive strength of the concrete may be utilized:

- for biaxial compression, $f_{2cd} = 1.20 f_{1cd} = f_{cd}$;
- for triaxial compression, $f_{3cd} = 3.88 f_{1cd} = 3.30 f_{cd}$.

When utilizing such high strengths a check must be made to confirm the magnitude of the transverse compression. Such high stresses usually occur only as local pressures, as for example under support plates. It is therefore very important that the flow of forces is further followed through the structure, because transverse tension may occur (see bottle in Fig. 5.8) requiring reinforcement.

More critical, and normally decisive in the design of a structure or member, is the anchorage of the ties. The anchorage of reinforcement requires transverse tension, which has to be taken by concrete in tension if there is no transverse reinforcement. The transfer of the forces into the struts should therefore be investigated thoroughly in three dimensions, e.g. both in the plane of load transfer and perpendicularly. The typical and often critical node is at an anchorage at the end support or at a concentrated load applied at the corner of a member, as shown in Fig. 5.9 for a bar with a hook. The anchorage length is normally defined from the inner face of the support and is calculated as an effective length $l_{b,net}$, as for example in Eurocode EC 2, Part 1:

$$l_{b,net} = \alpha_a l_b (A_{s,requ}/A_{s,prov})$$

where l_b is the basic anchorage length for a bar, and α_a is a factor for the anchorage element. Typical values for α_a are 1.0 for a straight bar, and 0.7 for a hook.

The favourable conditions for bond in such a node are due to the vertical compression and may be taken into account either by reducing the design value for the bond stress, or by reduction factors for the required anchorage length (calculated for a basic bond stress). The check on compressive stresses is often not critical at such nodes, because either the anchorage length or the limitation on bearing pressure tends to govern the dimensions of the node. For the effective strength in compression, a value $v_2 = 0.85$ may be used to allow for any transverse tension.[5.12]

In case of short anchorage lengths, the best detail for a node is of course provided by using anchor plates, because then the load transfer from the tie to the struts occurs as in a compression node. The anchor plate must be dimensioned for the assumed stress distribution at the face of the node.

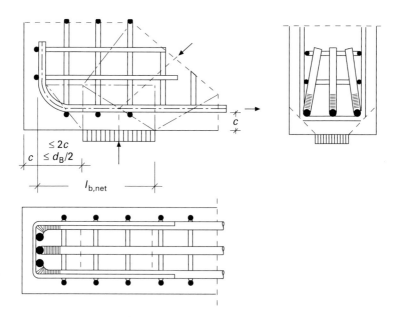

Fig. 5.9 Node at an end support of a beam with anchorage of a bar with a hook

5.4   Designing B- and D-Regions

5.4.1   Current Design Concepts

Current code design procedures concentrate on the dimensioning of cross-sections for the separate action effects (moment, shear etc.) and allow for interactions among the action effects by means of empirical combination rules (as for shear and torsion) and *ad hoc* detailing rules (such as the shift of the moment diagram line to allow for the effect of shear on the tensile force in beams).

Procedures for shear design exemplify the unsatisfactory state of the current design concepts,[5.38–5.40] and indeed have puzzled designers and researchers from the very beginnings of reinforced concrete. They appear in codes as empirical rules which contain ambiguities and even errors. Eurocode EC 2,[5.32] which is based on the CEB-FIP Model Code of 1978,[5.41] gives two different methods for shear design. The first is the older 'standard method', whereby the ultimate shear force is made up of a component V_{sw} carried by the stirrups, and a concrete term V_c:

$$V = V_{sw} + V_c \qquad [5.1]$$

The term V_{sw} represents truss action and is normally determined for a strut angle of $\theta = 45°$. The concrete term V_c is often set equal to the shear capacity of the beam in the absence of transverse reinforcement, although this is in principle wrong.[5.42–5.44]

The second method is the truss model with variable inclination of the struts, which is based on the theory of plasticity.[5.13,5.45–5.47] The transverse reinforcement is given by the equation:

$$V_d = V_{sw} = \frac{A_{sw}}{s_w} f_{ywd} z \cot \theta = (\rho_{sw} b_w) f_{ywd} z \cot \theta \qquad [5.2]$$

The angle θ of the compression field may be freely chosen, within certain limits. For example, the following limits have been proposed for the application of Eurocode EC 2 in Germany:

$$\tfrac{7}{4} > \cot \theta > \tfrac{4}{7} \quad \text{or} \quad 30° < \theta < 60° \qquad [5.3]$$

The existence of two alternative methods reflects differing views among researchers and code writers on shear design, and the designer is left without clear guidance for choosing between the two methods. The main criticism of the standard method is that the concrete contribution is determined empirically.[5.46] The main criticism of the truss method with variable angle θ is that the influence of axial forces and prestress is ignored. This may be unsafe in the case of axial tension, and conservative for high axial compression. Furthermore, a very low strut inclination has to be assumed for the design of members with small amounts of transverse reinforcement.

This confusion is surprising in view of the fact that truss models for shear design were proposed and used at a very early stage in the development of reinforced concrete. Unfortunately the further development of the early models did not occur until relatively recently.

5.4.2   Models and Theories for Shear Design

A unified view of the two alternative approaches for shear design is available.[5.48] The truss model with variable inclination of the struts is actually a *smeared truss model*, with basically a uniaxial compression field. The standard method may be regarded as a *discrete crack model*.[5.49–5.52] It is a failure mechanism approach because it considers conditions at the failure crack. As shown in Fig. 5.10, the shear force is transferred by various shear-carrying mechanisms. This leads to a consistent design method with a realistic assessment of the ultimate capacity.[5.49–5.53] An equation for the shear force follows from vertical equilibrium. In the case of a reinforced concrete member without prestressing tendons:

$$V = V_{sw} + V_f = \frac{A_{sw}}{s_w} f_{ywd} z \cot \beta_{cr} + V_f \qquad [5.4]$$

Here V_f is the shear force transferred by friction across the inclined crack surface (Fig. 5.10(b)), and β_{cr} is the angle of the inclined crack. The transverse reinforcement may thus be determined for a design shear force if β_{cr} and the contribution V_f are known. Fortunately simple expressions can be derived for V_f. For code purposes a constant value is a satisfactory approximation for reinforced concrete members.

When comparing Eqs 5.4 and 5.1 (for the 'standard method'), the similarity is obvious. The truss model with crack friction provides a physical explanation for the concrete contribution V_c. Once the value for V_f (Eq. 5.4) has been determined, the forces in the inclined section through the failure crack are known. The state of stress between the cracks can then be determined, and this is represented by a superposition of the two stress fields or truss models, as shown in Fig. 5.11.[5.52,5.53] The 'discrete crack approach' of Fig. 5.10 is therefore statically equivalent to the 'smeared' stress fields in Fig. 5.11, and hence this approach may also be called a 'truss model with discrete cracks' or a 'truss model with

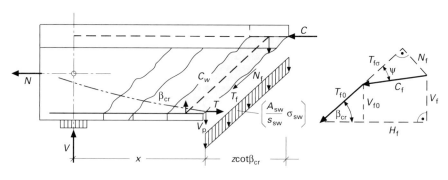

(a) Forces at the end support   (b) Forces from crack friction

Fig. 5.10 End support of a structural concrete beam with forces at the failure surface in the B-region[5.52,5.53]

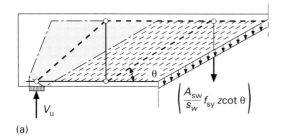

(a)

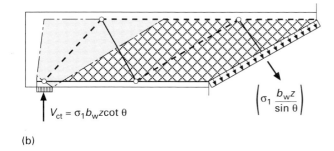

(b)

Fig. 5.11 Truss models and stress fields representing the state of stress and load transfer in cracked webs of structural concrete members[5.52,5.53]

crack friction'. These equivalences mean that discrepancies among the different code approaches can be physically explained and resolved.

The models in Fig. 5.11 are smeared truss models, in which the flow of forces is considered over the whole member, with the contribution to the truss by the concrete ties not being governed by the concrete tensile strength but by the friction forces in the crack. The models in Figs 5.11(a) and 5.11(b) are capable of dealing with the low-shear range as well as the transition to members without transverse reinforcement because the latter can be represented as in Fig. 5.11(b) using concrete tension fields or ties, as explained later in Section 5.5.

However, the two approaches differ considerably in their treatment of the strength of the compression struts. In the 'smeared' truss analogy the strength of the compression strut is limited to an 'effective strength' $f_{cw,eff}$, which covers all the influences listed in Section 5.3.2. Of special importance for webs is the reduction in strength due to stress transfer over cracks, and according to EC 2, Part 1, $f_{cw,eff}$ is limited to $0.50f_{cd}$ for concrete strengths higher than $f_{ck} = 40$ MPa. In the 'truss model with crack friction' the compression failure in the web only occurs with high shear, where the struts are almost parallel to the cracks; hence the higher value of $0.80 f_{1cd}$ was used above. This leads to differences between the two approaches as demonstrated in Fig. 5.12, which gives the required amount of transverse reinforcement versus the ultimate shear force for the web of a reinforced concrete beam. Large discrepancies between these methods can be seen in Fig. 5.12 in the case of medium shear, particularly with respect to the required transverse reinforcement, and also for high shear, with respect to the maximum allowable shear force.

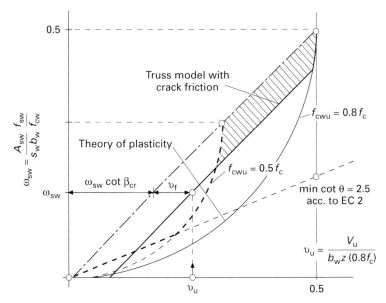

Fig. 5.12 Transverse web reinforcement for a reinforced concrete beam according to the truss model with crack friction and the truss model with variable strut angle (EC 2)[5.32]

This brief review shows that the model for structural behaviour in B-regions is clear and unambiguous. In the following examples, clear results are obtained from the truss model, while the code methods can lead to uncertainty. For simplicity, the cases treated are for high shear forces, so that the contribution of the tension truss in Fig. 5.11(b) is small and can be neglected. That is to say, only the model with a uniaxial compression field (Fig. 5.11(a)) is used.

5.4.3   Design of B-Regions

Geometry of the Truss Model
For practical design it is first necessary to establish the geometry of the truss. This is defined by the distance between the chords (the inner lever arm) and the inclination of the compression struts. The magnitudes of these values follow from the well-known classical methods for the design of structural concrete, i.e. flexural design and shear design. The inner lever arm z is obtained from the flexural design at sections of maximum moment, and mostly the minimum value is applied throughout the length of the beam. The following approximate values are often used:

- for rectangular cross-sections,

$$z = 0.9d \qquad\qquad [5.5a]$$

- for T-beams or box beams,

$$z = d - \frac{h_f}{2} \tag{5.5b}$$

The strut inclination is either directly given or may be freely selected (see Eq. 5.3). In the case of the 'truss model with crack friction' (see Section 5.4.2) it must be evaluated by first setting Eq. 5.4 equal to Eq. 5.2:

$$V_{sw} = V_d - V_f = (\rho_{sw} b_w) f_{ywd} z \cot \theta$$

This leads to:

$$\cot \theta = \frac{V_d - V_f}{(\rho_{sw} b_w) f_{ywd} z} \tag{5.6}$$

The same may be done for Eq. 5.1 for the 'standard method', so that in Eq. 5.6 the term V_c replaces the term V_f. In a non-dimensional design diagram such as Fig. 5.12, Eq. 5.6 is represented by a straight line from the origin to the relevant design point as shown in Fig. 5.13.

Forces in the Truss Model (B-Regions)
The forces in the stirrups and chords in the B-region are best derived by considering an inclined section parallel to the struts (Fig. 5.14). Vertical equilibrium leads to Eq. 5.2, and also to the following expression for stirrup force per unit length:

$$n_{sw} = \frac{A_{sw}}{s_w} f_{ywd} = \frac{V}{z \cot \theta} \tag{5.7}$$

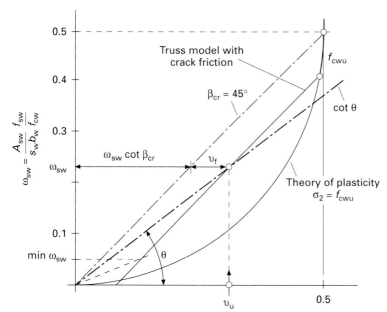

Fig. 5.13 Comparison of fictitious strut angles for shear design

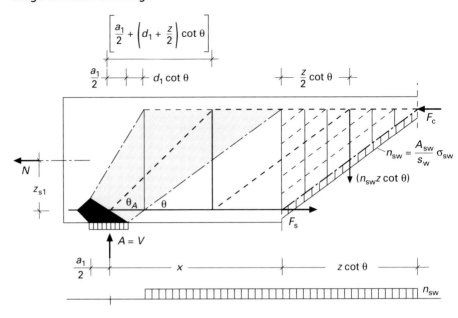

Fig.5.14 Forces in an inclined section parallel to the struts of a truss

For a design diagram like the one in Fig. 5.13, Eq. 5.2 is divided by $b_w z f_c$, so that the ultimate shear force may be related to a mechanical reinforcing ratio ω_{sw} (as in flexural design):

$$v_u = \frac{V_u}{b_w z f_{cwu}} = \rho_{sw}\frac{f_{ywd}}{f_c}\cot\theta = \omega_{sw}\cot\theta \qquad [5.8]$$

with

$$\omega_{sw} = \rho_{sw}f_{yw}/f_{cwu} \quad \text{and} \quad \rho_{sw} = A_{sw}/(b_w s_w) \qquad [5.9]$$

$$v_u = \tau_{0u}/f_{cwu} \quad \text{and} \quad \tau_{0u} = V_u/b_w z \qquad [5.10]$$

The non-dimensional values v and ω may of course also refer to other strength values such as f_c (vf_{cd}). The tension chord force is

$$F_s = \frac{M}{z} + \frac{V}{2}\cot\theta + N\left(1 - \frac{z_{s1}}{z}\right) \qquad [5.11]$$

The force in the compression chord is

$$F_c = \frac{M}{z} - \frac{1}{2}V\cot\theta + N\frac{z_{s1}}{z} \qquad [5.12]$$

The stress σ_{cw} in the inclined struts follows from a vertical section:

$$\sigma_{cw} = \frac{F_{cw}}{b_w z \cos\theta} = \frac{V}{b_w z}\frac{1}{\sin\theta\cos\theta} \qquad [5.13]$$

and non-dimensionally:

$$\frac{\sigma_{cw}}{f_{cwu}} = \frac{V}{b_w z f_{cwu}} \frac{1}{\sin\theta\cos\theta} = v\frac{1}{\sin\theta\cos\theta} \tag{5.14}$$

Now all the forces and stresses are known in the B-region and the design may be carried out. It will be seen in the following that the model of the B-region must be known before the modelling of the adjacent D-regions can be carried out.

5.4.4   Static Discontinuities at Point Loads

End Support of a Beam

At an end support the compression field in the web must change to a fan (see Fig. 5.3(a)), so that it can equilibrate the concentrated support force and the force in the tension chord at the node (Fig. 5.9). The geometry of the fan may be defined in various ways. In Fig. 5.15 the fan starts where the parallel compression field of the B-region first meets the inner face of the support. The first stirrup in the B-region is consequently placed at a distance $d_1\cot\theta$ from the inner face of the support, where the boundary line between the fan and the parallel compression stress field, i.e. between the B- and D-regions, intersects the line of the tension chord. The maximum angle of the fan (i.e. of the outer boundary line) is defined by the intersection of the first stirrup with the axis of the compression chord. The angle θ_A for the resultant strut in the fan follows from the geometry:

$$\cot\theta_A = \left[\frac{1}{2}\frac{a_1}{z} + \left(\frac{d_1}{z} + \frac{1}{2}\right)\cot\theta\right] \tag{5.15}$$

From this we find that the following force in the tension chord has to be anchored in the node:

$$F_s = A\cot\theta_A \tag{5.16a}$$

or in case of an additional axial force:

$$F_s = A\cot\theta_A + N\left(1 - \frac{z_{s1}}{z}\right) \tag{5.16b}$$

For typical dimensions, calculations in Fig. 5.15 for a strut angle of 30° ($\cot\theta = 1.75$) give a value of $F_s = 1.21A$ for the tension force to be anchored at the support. This is larger than the value given in some codes; in particular, Section 5.4.2.1.3 in EC 2 is unsafe:

$$F_s = Va_1/d$$

where a_1 is the horizontal shift of the M/z line. Applying this to the 'method with variable strut inclination' the shift is

$$a_1 = \frac{z\cot\theta}{2}$$

and this leads to a force of $F_s = 0.79A$, which is only about two-thirds of the value obtained above from the truss model. This emphasizes the importance of applying a model for the design of crucial details in D-regions such as the end support.

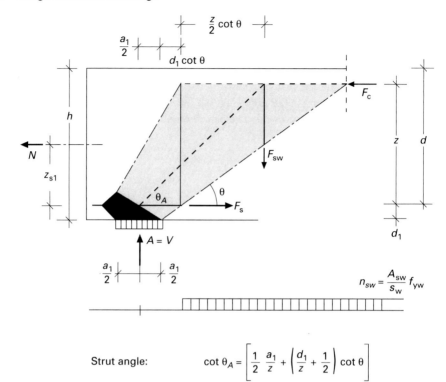

Strut angle: $\cot\theta_A = \left[\dfrac{1}{2}\dfrac{a_1}{z} + \left(\dfrac{d_1}{z} + \dfrac{1}{2}\right)\cot\theta \right]$

Force in tension chord at support: $F_s = A\cot\theta_A = V\cot\theta_A$

Example: $a_1 = 0.2h;$ $d = 0.9h;$ $z = 0.9d$

$\cot\theta_A = 0.124 + 0.623\cot\theta$

$\cot\theta = 1.75:\ \dfrac{F_s}{A} = \cot\theta_A = 1.21$

Fig. 5.15 Strut-and-tie model and stress field at an end support

Beam with an Intermediate Support

An intermediate support contains two fans in opposite directions and can simply be regarded as a combination of two end supports for the relevant shear forces, which are components of the total support force. In Fig. 5.16 two point loads are applied to the beam and the distribution of the forces in the tension chord and in the web are plotted over the entire length. The ratio of the point load in the span to that on the cantilever has been selected to produce equal moments of $M = 4Fz$ at mid span and over the support.

The region at the end support corresponds to Fig. 5.15 except that the support force here is different. For the cantilever, the same model also applies, although a steeper angle θ for the inclined compression field has been selected because of the higher shear force of $1.0F$ instead of $0.67F$ at the end support. It should be noted that the node on the compression chord under the in-span load is not

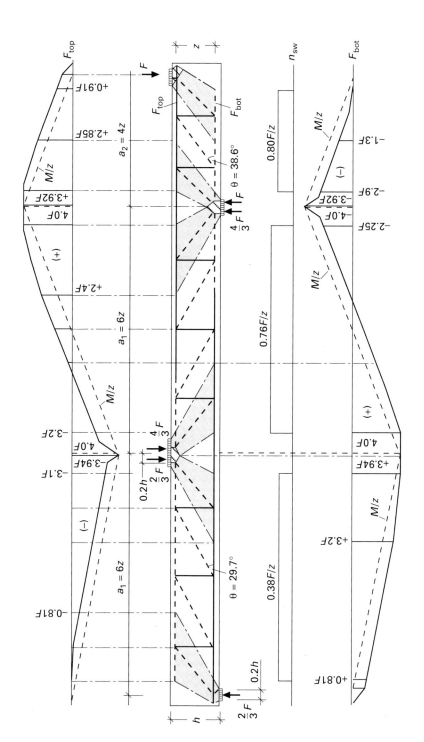

Fig. 5.16 Strut-and-tie model and stress fields for a beam with an intermediate support

symmetric; neither is the node above the intermediate support. In each case the width of the load or support plate has been subdivided into two parts according to the proportion of the applied loads (the shear forces). In the region between the in-span load and the intermediate support the section with zero moment is of some interest because the tension forces in the top and bottom chords overlap over an extended region.

5.4.5   A Common Misunderstanding of Modelling Beams with Strut-and-Tie Models

The terms *strut-and-tie model* and *truss model* are often misunderstood, in that it is thought that a complete truss with discrete struts always must be found for a given beam geometry. This is the case in Fig. 5.16 above, but only because the lengths a_1 and a_2 were adjusted for didactic reasons to the selected strut angles, so that the load transfer could be easily visualized. Generally a complete truss cannot be expected because there are different load cases and the magnitude of the shear varies, so that different strut angles result for a given beam geometry. Figure 5.17(a) shows such a case in which the struts of the truss model do not 'meet' in the span, when modelling is started with a given strut angle from the two opposite load points. This means that the length a_B of the B-region between the borders of the two fans of the two D-regions is arbitrary and not a multiple of the truss length ($z\cot\theta$). This however is not a problem. Figures 5.17(b) and 5.17(c) show two different ways for deriving the length such that the B-region may be of any length and the web is always in equilibrium.

5.4.6   Beams with Distributed Loading

The model in the web of a beam with distributed loading consists of elements which have previously been discussed: a parallel uniaxial compression stress field and a fan-shaped stress field at the end support, as shown in Fig. 5.18. For simplicity the inner lever arm has been set to 1.0 m, so that the truss length for a strut angle of about 30° is ($z\cot\theta$) = 1.75 m. The loading on the top chord in the region of the fan is transferred directly to the support; the transverse reinforcement near the end support has only to 'hang' the loading beyond the inclined section 1–1 defining the end of the fan or of the D-region. The first option for the distribution of the transverse reinforcement is a linear one, which corresponds to the distribution of the shear force (Fig. 5.18). The required transverse reinforcement in the B-region has to be designed for the following force parameter:

$$v_{sw} = \frac{V}{z\cot\theta} - \frac{p}{2} \qquad [5.17]$$

In any sectional design, the shear force may therefore be taken to be slightly less than the actual shear force in that section. This distribution of the transverse tension force in the web leaves a certain length free from transverse reinforcement near mid-span, which is equal to half the truss length. Because of this, another fan-shaped compression stress field appears in the web. It may be noted that it is

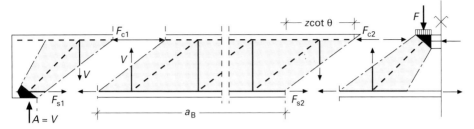

(a) B-region of arbitrary length a_B

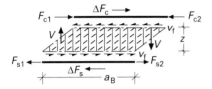

$\Sigma H = 0: \Delta F_c = F_s = v_f a_B$

$\Sigma M = 0: (v_f a_B)z = V a_B$

$$\boxed{v_f = \frac{V}{z}} \text{ or with } v_f = \tau_0 \, b_w, \ \tau_0 = \frac{V}{b_w \, z}$$

(b) Equilibrium of B-region

$n_{xw} = \dfrac{V\cot\theta}{z}$

$\Sigma M = 0: (v_f a_B)z = (v_w z)a_B$

$$\boxed{v_f = v_w = \frac{V}{z}}$$

$v_w = V/z$

(c) Equivalent consideration as in (b)

Fig. 5.17 Equilibrium for a B-region of arbitrary length

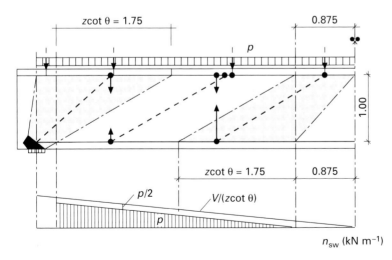

Fig. 5.18 Stress fields for a beam with distributed loading, linearly distributed stirrups

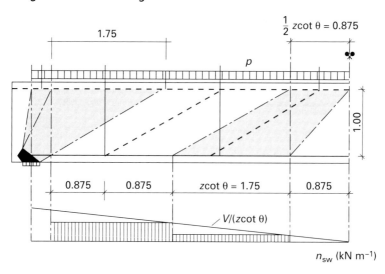

Fig. 5.19 Truss model and stress fields for a beam with distributed loading, staggered stirrups

not possible to show a complete truss over the beam length. The reason is that the resultants of the stirrup forces are not in the middle of the stress fields, due to the linear distribution of the transverse reinforcement, whereas the resultants of the loadings are. The second and more practical option is to stagger the transverse reinforcement as shown in Fig. 5.19. Marti has shown that this fully complies with the theory of plasticity.[5.54] As mentioned before, it is pure chance that a complete truss occurs in this figure.

5.4.7   Indirectly Loaded or Supported Beams

The term 'indirect loading' means that the load is not applied directly to the top compressive surface of the member, but where two members intersect, with the load transfer causing vertical tension. Figure 5.20 shows an end support of a beam (I), which is supported by another beam (II). The load transfer in beam I does not differ from that of a 'directly' supported beam, as can be seen by comparison with Fig. 5.15. However, the total support force has to be 'hung' by reinforcement to the top of beam II, where it is again transferred into the beam by two fan-shaped compressive stress fields.

The main elements of the models reappear. The main difference is that the fan of beam I has to transfer its load to the 'hanging' reinforcement within the intersection of both beams. The model clearly shows that, contrary to some codes, any stirrups outside the intersection cannot be recognized as hanging reinforcement. The model also makes it obvious that the stirrups for this reinforcement have to placed within beam I in order to provide a wide support for the fan in beam I. The load transfer to the top of beam II is provided by the two legs of the hanging reinforcement, so that a concentrated node occurs there. This cannot be

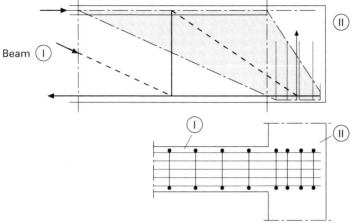

(a) Indirect end support and arrangement of stirrups

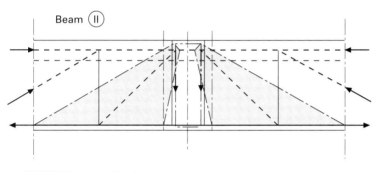

(b) STM in supporting beam

Fig. 5.20 Indirect end support for a beam

avoided, but it is better here in the compression zone than in the bottom of the intersection, where the longitudinal reinforcement of beam I has to be anchored. This anchorage is very unfavourable because of the transverse tension, and this should be recognized when calculating the required anchorage length.

Altogether this example shows that the designer can decide on the best solution for detailing without having to rely on code regulations, if there are any. Furthermore, the close interaction between detailing, dimensioning and modelling is obvious. The traditional separation of dimensioning and detailing into different chapters of codes is thus dubious.

5.4.8   Point Load Near an End Support

The design for a point load near an end support has long been a controversial topic in codes, and has not usually been adequately dealt with. The reason is probably that it has been seen as a shear problem, whereas it has to be treated as a complete D-region, like a deep beam. The basic idea is clear: for a point load F acting closer than about $a < 2d$ to a support, the transverse reinforcement need not be designed for the full shear force because some part of the load (F_2) may be transferred directly to the support by an inclined strut, as can be seen from the strut-and-tie model and stress fields in Fig. 5.21. The other part of the load F_1 is transferred into the web by means of a fan-shaped compression field and has to be 'hung' from the compression zone by stirrups.

A limiting length a_{lim} must first be determined for which the model in Fig. 5.21 remains valid. With regard to Fig. 5.17 it was explained that the B-region can be of any length. If its length is zero, the fans of the D-regions just intersect, as shown in Fig. 5.22(a). This is the limiting case, where the total load F is carried by stirrups placed between the load and the support. However, it is obvious that the limiting distance of the load from the support line is not a fixed value, but

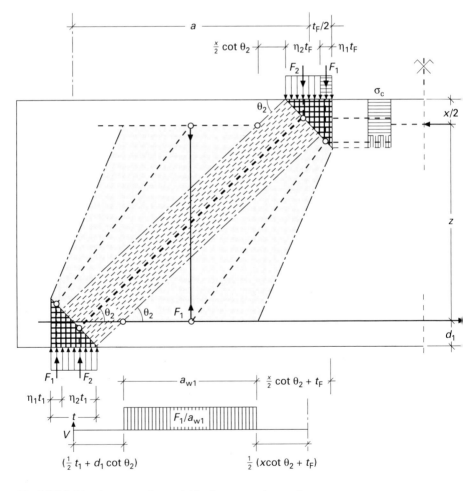

Fig. 5.21 Model and geometry for a point load near an end support

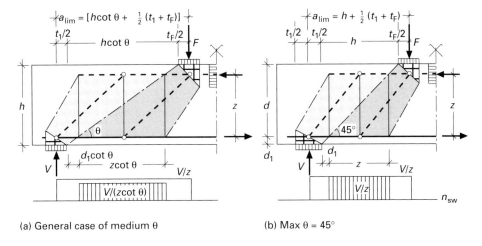

(a) General case of medium θ (b) Max $\theta = 45°$

Fig. 5.22 Model for point load at $a_{\lim}$ from the support, with no B-region

rather depends very much on the angle of the struts in the web. According to Fig. 5.22(a) the value for $a_{\lim}$ is

$$a_{\lim} = [0.5(t_1 + t_F) + h\cot\theta] \qquad [5.18]$$

where t_1 and t_F are the widths of the loading plates. For a strut angle of about 30° (cot $\theta = 1.75$) and support widths of $t_1 = t_F = 0.1h$ the value $a_{\lim} = 1.85h$ is obtained, which is equal to $a_{\lim} = 2.06d$ for $d = 0.9h$. This agrees well with the value usually assumed, i.e. $2d$. Greater values than these should not be used because they require unrealistically small values for θ. However, with steeper strut angles smaller values are obtained for $a_{\lim}$. The limiting distance for an extreme case is shown in Fig. 5.22(b), where the web is fully utilized in compression, so that the strut angle is 45°. Then $a_{\lim} = 1.1h$ or $a_{\lim} = 1.22d$. This is far smaller than $2d$, which is usually assumed. Hence there is a safety risk for highly utilized webs. The model in Fig. 5.21 is internally statically indeterminate because there are two load paths. There are various proposals for determining the magnitude of the part F_1 of the load. The following relationship,

$$\frac{F_1}{F} = \frac{1}{2}\left(\frac{3a}{a_{\lim}} - 1\right) \qquad [5.19]$$

was derived empirically from tests on beams without stirrups.[5.52] For distances $a_{\min} < a_{\lim}/3$ the load component F_1 is zero and no transverse reinforcement is required. For the above calculated value of $2.06d$ for $\theta = 30°$, this distance is $a_{\min} = 0.68d$ (or $a_{\min} = 0.76z$ for $z = 0.9d$), which corresponds to the well-known strut-and-tie model for corbels. For loads very near to the supports design as a corbel may not be adequate. Figure 5.23 demonstrates that the inclined strut may split and that horizontal reinforcement should be provided in corbels.

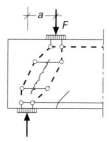

Fig. 5.23 Possible splitting of strut for a point load close to an end support

5.4.9   Model for the Design of Columns

The design of columns requires two steps: first the additional moments due to the deformations have to be assessed (buckling design), and then the dimensioning for the total moment may be carried out as for any other member. Figure 5.24 with a pin-ended column shows that in principle the model for columns is the same as that already considered for beams. The main differences are that the angle of the inclined compressive stress field is flatter and that the chords are uncracked over a considerable length. The web is also uncracked in the lower part of the column shown in Fig. 5.24 because the cross-section is rectangular. The same model may be used for a prestressed beam, but then the web may be cracked in the case of an I-section, although the chords may be uncracked.

5.5   Design Model for Members Without Transverse Reinforcement

Current design procedures for members without shear reinforcement are based on empirical formulae which do not provide the designer with any understanding of structural behaviour. Design is thus reduced to checking code rules which cannot deal with new or unusual design problems.

The structural behaviour of slender members without shear reinforcement can only be explained by the 'tooth-model', which was originally proposed by Kani,[5.55,5.56] and further developed by others.[5.57–5.59] An important and relevant fact which has been well known since Kani's tests is that the strut-and-tie model is unsafe according to the theory of plasticity (the tied arch model for members with distributed loads). This has been substantiated by more recent investigations.[5.60,5.61]

Based on previous work, Reineck[5.52,5.53] has developed a mechanical model for beams without transverse reinforcement, taking account of the crack pattern, with concrete teeth between the cracks. Failure is governed by the shear transfer mechanisms at the discrete cracks, which include friction at the crack faces (aggregate interlock for normal-weight concretes) and dowel action in the longitudinal reinforcement. Following clear mechanical principles, Reineck has derived an explicit equation for the ultimate shear force of members without transverse reinforcement. The proposed model considers all the well-known parameters influencing shear capacity, such as the depth of the member (size effect) and the reinforcement ratio. For practical applications a dimensioning

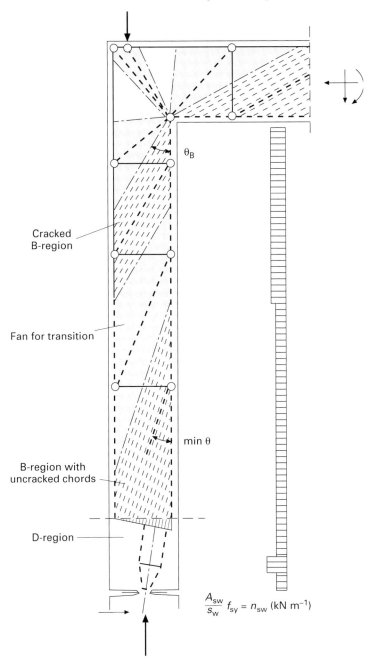

Figure 5.24 Strut-and-tie model and stress fields for a column

diagram is available for reinforced concrete members, but the theory also gives a consistent treatment of axial forces.

In this theory the state of stress in the member is clearly defined in terms of forces at the discrete cracks. The principal stresses between the cracks may also be resolved, which are essentially an inclined biaxial tension–compression field in the concrete, due to the friction stresses along the crack. This may be visualized using the same truss model that was shown in Fig. 5.11, where the tension field in the concrete is perpendicular to the compression field, which is itself inclined at 30°. This model thus provides a consistent transition from members without transverse reinforcement to members with minimum and low amounts of transverse reinforcement. This is a very important feature for structural concrete design and is of practical relevance for many structures.

This model with concrete ties clearly demonstrates that the concrete tensile strength is actually utilized in reinforced concrete members without transverse reinforcement. Furthermore, in regard to the practical aspect of dimensioning and detailing, the forces in the tension chord may be quickly determined from the truss analogy in order to stagger the longitudinal reinforcement. It has also been demonstrated using practical examples that strut-and-tie modelling can be extended to B- and D-regions without transverse reinforcement.[5.48,5.52]

5.6   Summary and Conclusions

It is now possible to carry out the design of structural concrete in a consistent manner using strut-and-tie and stress-field models. Such models can be used to dimension regions with statical or geometric discontinuities (D-regions). These are poorly treated in current standards, and are frequently the source of damage and structural failure. The B-regions are of course covered by truss models, and these have been used for many years.

A brief account has been given in this chapter of the strut-and-tie approach, and of various ways of developing appropriate models. The strengths of the basic elements of the strut-and-tie models have been considered. Although the strength of concrete struts has been considered in some detail, in most design situations the simple values are sufficient. In well-designed structures, strength (and therefore the design) is not governed by concrete in compression, but by yielding of the reinforcement. Of considerable importance is the design of the nodes, particularly those where reinforcement is anchored. The strut-and-tie approach draws attention to these critical regions, as well as allowing an overall visualization of the flow of forces through the structure.

References

Abbreviations used:

ACI: American Concrete Institute
ASCE: American Society of Civil Engineering
BuStb: Beton- und Stahlbetonbau

CEB: Comité Eurointernational du Béton
DAfStb: Deutscher Ausschuss für Stahlbeton
FIP: Federation Internationale Precontrainte
IABSE: International Association for Bridge and Structural Engineering
PCI: Prestressed Concrete Institute

5.1   Ritter W 1899 Die Bauweise Hennebique. *Schweizerische Bauzeitung* Bd XXXIII,
       Nr 7, Jan (The Hennebique method)
5.2   Mörsch E 1912 *Der Eisenbetonbau* 4 Aufl, K Wittwer, Stuttgart (*Reinforced
       Concrete Construction* 4th edn)
5.3   Mörsch E 1929 *Der Eisenbetonbau* 6 Aufl, I Bd, 2 Hälfte, K Wittwer, Stuttgart
       (*Reinforced Concrete Construction* 6th edn)
5.4   Leonhardt F, Lippoth W 1970 Folgerungen aus Schaden an Spannbetonbrücken.
       BuStb **65**(10): 231–244 (Conclusions to be drawn from damage observed in
       prestressed concrete bridges)
5.5   Leonhardt F, Koch R, Rostasy F S 1971 Aufhängebewehrung bei indirekter
       Lasteintragung von Spannbetonträgern, Versuchsbericht und Empfehlungen. *BuStb*
       66(10): 233–241. Zuschrift von Baumann T 1972 *BuStb* **67**(10): 238–239
       (Hanging reinforcement for the indirect introduction of loads in prestressed
       concrete girders)
5.6   Podolny W 1985 The cause of cracking in post-tensioned concrete box girder
       bridges and retrofit procedures. *PCI Journal* **30**(2): 82–139
5.7   Jakobsen B 1992 The loss of the Sleipner A Platform, *Proceedings Second
       International Offshore and Polar Engineering Conference* San Francisco, USA, 14–
       19 June, Vol 1, pp 1–8
5.8   Wagner P 1992 Der Untergang der Plattform Sleipner A. *Bautechnik* **69**(8): 449–
       450 (The sinking of the Sleipner Platform)
5.9   Schlaich J, Reineck K-H 1993 Die Ursache für den Totalverlust der Betonplattform
       Sleipner A. *BuStb* **88**(1): 1–4 (The cause of the loss of the concrete Offshore
       Platform Sleipner A)
5.10   Schlaich J 1984 Zur einheitlichen Bemessung von Stahlbetontragwerken. *BuStb* **79**:
       89–96 (Consistent design of concrete structures)
5.11   Schlaich J, Schäfer K, Jennewein M 1987 Toward a consistent design for structural
       concrete. *PCI Journal* **32**(3): 75–150
5.12   Schlaich J, Schäfer K 1993 Konstruieren im Stahlbetonbau. *Beton-Kalender* Teil II,
       pp 327–486, W Ernst u Sohn, Berlin (Designing in structural concrete)
5.13   Thürlimann B, Grob J, Lüchinger P 1975 *Torsion, Biegung und Schub in
       Stahlbetonträgern* Institute for Structural Engineering, ETH Zürich, Report 170
       (*Torsion, Flexure and Shear in Reinforced Concrete Girders*)
5.14   Thürlimann B, Marti P, Pralong J, Ritz P, Zimmerli B 1983 Anwendung der
       Plastizitätstheorie auf Stahlbeton. *Fortbildungskurs für Bauingenieure*, Institut für
       Baustatik und Konstruktion, ETH Zürich (Application of plasticity theory to
       reinforced concrete)
5.15   Thürlimann B 1985 *Plastizitätstheorie im Stahlbetonbau* Lecture series, summer
       term, Institut für Massivbau, Univ Stuttgart (*Theory of Plasticity in Reinforced
       Concrete*)
5.16   Marti P 1985 Basic tools of reinforced concrete beam design. *ACI Journal* **82**(1):
       46–56; Discussion, **82**(6): 933–935
5.17   Stone W C, Breen J E 1984 Design of post-tensioned girder anchorage zones. *PCI
       Journal* **29**(1): 64–109, and **29**(2): 28–61

5.18   Collins M P, Mitchell D 1980 Shear and torsion design of prestressed and non-prestressed concrete beams. *PCI Journal* **25**(5): 32–100; Discussion, 1981 **26**(6): 96–118

5.19   Collins M P, Mitchell D 1986 A rational approach to shear design – the 1984 Canadian code provisions. *ACI Journal* **83**(6): 925–933

5.20   Collins M P, Mitchell D 1987 *Prestressed Concrete Basics* Canadian Prestressed Concrete Institute, Ottawa

5.21   Leonhardt F 1965 Über die Kunst des Bewehrens von Stahlbetontragwerken. *BuStb* **60**(8): 181–192; (9): 212–220 (On the art of detailing in reinforced concrete)

5.22   Park R, Paulay T 1975 *Reinforced Concrete Structures* John Wiley and Sons, New York

5.23   Paulay T 1989 Equilibrium criteria for reinforced concrete beam-column joints. *ACI Structural Journal* **86** (Nov–Dec): 635–643

5.24   IABSE 1991 Proceedings of Colloquium on Structural Concrete, Stuttgart. *IABSE Report* **62**: 1–872.

5.25   IABSE 1991 IABSE Colloquium, summarising statement. *PCI Journal* **36** (Nov–Dec): 60–63

5.26   Breen J E 1991 Why structural concrete?, Ref. 5.24, pp 15–26

5.27   Schlaich J 1991 The need for consistent and translucent models. Ref. 5.24, pp 169–184

5.28   Schleeh W 1964 Ein einfaches Verfahren zur Lösung von Scheibenaufgaben. Sonderdruck aus *BuStb* **59** (3, 4, 5): 1–20 (A simple procedure for solving the plate problem)

5.29   Nielsen M P, Braestrup M W, Jensen B C, Bach F 1978 *Concrete Plasticity: Beam Shear, Shear in Joints, Punching Shear* Danish Society for Structural Science and Engineering

5.30   Muttoni A, Schwartz J, Thürlimann B 1988 *Bemessen und Konstruieren von Stahlbetontragwerken mit Spannungsfelden* Vorlesungen SS, Institut für Baustatik und Konstruktion, ETH Zürich (*Design of Reinforced Concrete Structures with Stress Fields*)

5.31   Rückert K 1994 Computer-unterstützes Bemessen mit Stabwerkmodellen (*Computer-based proportioning, using strut-and-tie methods*). *BuStb* **89**(12): 319–325

5.32   Eurocode 2 1991–1992 Planung und Bemessung von Stahlbeton- und Stahlbetontragwerken, 1. *Beton-Kalender* Teil II B, pp 681–815 (Design of reinforced concrete and prestressed concrete structures)

5.33   CEB-FIP 1993 *Model Code 1990* MC 90, Thomas Telford

5.34   Schäfer K, Schelling G, Küchler T 1990 Druck- und Querzug in bewehrten Betonelementen *DAfstb* H408, Beuth Verlag, Berlin (Compression and transverse tension in reinforced concrete elements)

5.35   Eibl J, Neuroth U 1988 *Untersuchungen zur Druckfestigkeit von bewehrtem Beton bei gleichzeitig wirkendem Querzug* Institut für Massivbau und Baustofftechnologie, Univ Karlsruhe (*Investigation into the Compressive Strength of Reinforced Concrete with Simultaneous Transverse Tension*)

5.36   Kolleger J, Mehlhorn G 1990 Experimentelle Untersuchungen zur Bestimmung der Druckfestigkeit. *DAfStb* H 413, Beuth Verlag, Berlin (Experimental investigation to determine compressive strength)

5.37   Reineck K-H 1989 Theoretical considerations and experimental evidence on web compression failures of high strength concrete beams. *CEB Bulletin* **193**: 61–73

5.38   MacGregor J G 1984 Challenges and changes in the design of concrete structures. *Concrete International* (Feb): 49–53

5.39 MacGregor J G 1991 Dimensioning and detailing. In *IABSE-Colloquium Stuttgart 1991: Structural Concrete. IABSE Report* **62** 391–409. Zurich

5.40 Marti P 1991 Dimensioning and detailing. In *IABSE-Colloquium Stuttgart 1991: Structural Concrete Report* **62** 411–443. Zurich

5.41 CEB-FIP 1978 *Mustervorschrift für Tragwerke aus Stahlbeton und Spannbeton* MC 78, Richtlinien, 3 Ausg (*Model Code for Reinforced and Prestressed Concrete Construction*)

5.42 Leonhardt F 1965 Die verminderte Schubdeckung bei Stahlbetontragwerken. *Der Bauingenieur* **40**(1): 1–15 (Reduced shear cover for reinforced concrete construction)

5.43 Leonhardt F 1977 Schub bei Stahlbeton und Spannbeton, Grundlagen der neueren Schubbemessung. *BuStb* **72**(11): 270–277; **72**(12): 295–302 (Shear in reinforced concrete and prestressed concrete)

5.44 Leonhardt F 1978 Shear in concrete structures. *CEB Bulletin* No 126: 67–124

5.45 Thürlimann B 1978 Shear strength of reinforced and prestressed concrete beams. *CEB Bulletin* No 126: 16–38

5.46 Thürlimann B 1979 Shear strength of reinforced and prestressed concrete beams, CEB approach. *Concrete Design, US and European Practices, ACI* **SP-59**: 93–115

5.47 Braestrup M W 1974 Plastic analysis of shear in reinforced concrete. *Magazine of Concrete Research* **26**(89): 221–228

5.48 Reineck K-H 1995 Shear design based on truss models with crack friction. *CEB Bulletin* No 223: 137–157.

5.49 dei Poli S, Gambarova P S, Karakoe C 1987 Aggregate interlock role in reinforced concrete thin-webbed beams in shear. *ASCE Journal of Structural Division* **113**(1): 1–19

5.50 Kupfer H, Mang R, Karavesyrouglou M 1983 Failure of the shear zone of reinforced concrete and prestressed concrete girders, an analysis with consideration of interlocking of cracks (in German). *Bauingenieur* **58**: 143–149

5.51 Kupfer H, Bulicek H 1991 Comparison of fixed and rotating crack models in shear design of slender concrete beams. *Progress in Structural Engineering* (ed Grierson D E *et al*) Kluwer, Brascia, Italy, pp 129–138

5.52 Reineck K-H 1990 Mechanical model for the behaviour of reinforced concrete members in shear. Dr Thesis, Univ Stuttgart

5.53 Reineck K-H 1991 Modelling of members with transverse reinforcement. Ref. 5.24, 481–488

5.54 Marti P 1986 Staggered shear design of simply supported concrete beams. *ACI Journal* **83**(1): 36–42

5.55 Kani G N J 1964 The riddle of shear failure and its solution. *ACI Journal* **61**(4): 441–467; Discussion, 1964, **61**: 1587–1636

5.56 Kani G N J 1966 Basic facts concerning shear failure *ACI Journal* **63**: 1518–1528

5.57 Fenwick R C, Paulay T 1968 Mechanisms of shear resistance of concrete beams. *ASCE Journal* **94**(ST 10): 2325–2350

5.58 Taylor H J P 1974 The fundamental behaviour of reinforced concrete beams in bending and shear. *ACI* **SP-42**: 43–77

5.59 Hamadi Y D, Regan P E 1980 Behaviour in shear of beams with flexural cracks. *Magazine of Concrete Research* **32**(1): 67–77

5.60 Reineck K-H 1991 Ultimate shear force of structural concrete members without transverse reinforcement derived from a mechanical model. *ACI Structural Journal* **88**(5): 592–602

5.61   Muttoni A 1990 Die Anwendung der Plastizitätstheorie in der Bemessung von Stahlbeton. Dissertation, Inst für Baustatik und Konstruktion, ETH Zürich (Use of the theory of plasticity for dimensioning reinforced concrete)

6 Fire Engineering Considerations

Karl Kordina and Marita Kersken-Bradley

Fire protection strategies need to be considered at a very early stage in the design of new buildings, and also in the renovation of existing buildings. Early planning can result in very significant cost savings. On the other hand, costs can become very high, and indeed impossibly high, if consideration is first given to fire safety when the building work is nearing completion.

Fire protection is achieved through a range of measures, including planning of evacuation routes and protection against smoke and toxic gases, as well as the fire design of loadbearing members.

Specific fire design provisions depend on the type and function of the building, for example whether a high-rise office building or a large industrial complex. They also vary according to local and national standards. Attention is therefore focused in this chapter on general aspects of fire safety in large concrete buildings.

6.1 Fire Risks

6.1.1 Objectives and Strategies

Fires in buildings occur with unwelcome frequency in industrialized countries, and despite the efforts of safety authorities, fire frequency has not been substantially reduced in recent years. Designers have a very important role to play in providing adequate fire safety in buildings. Through the appropriate choice of design details and materials, they must protect lives, minimize damage and economic loss, and employ measures which restrict the spread of fire.[6.1] Modern structures with finely proportioned, slender and highly stressed members can pose particular problems to designers with regard to achieving good fire safety.

National regulations for fire protection in buildings have a variety of aims, which can be summarized as follows:

- prevention of the outbreak of fire;
- prevention of the spread of fire, smoke, hot and toxic gases;
- safety of occupants in the fire zone, in other parts of the building and in adjacent buildings;
- evacuation of occupants from the immediate fire zone and other parts of the building;
- mitigation of damage to the building structure, the contents and to adjacent buildings;
- protection of life during firefighting and rescue operations.

Government agencies are primarily interested in life safety, which in turn emphasizes measures aimed at fire containment, and delaying the growth and spread of fire. They are also concerned with the possibility of general uncontrolled conflagration in built-up areas, and in ensuring that firefighting and rescue operations can proceed, even in buildings with difficult access such as large or high buildings. It is becoming apparent that some industrial activities also require national regulation because of special hazards which occur in the event of malfunction or accident. Examples are fires in nuclear power stations and in underwater tunnels. Thus, government agencies need to consider a variety of aspects in addition to life safety when developing requirements for fire protection. On the other hand, the mitigation of property damage is usually looked after by insurance activities on a voluntary rather than a mandatory basis.

Overall fire protection strategies involve both active and passive (or built-in) measures.[6.2] The passive built-in measures are preventive in character and remain permanently effective. Active measures, on the other hand, require the stimulus of the fire. They include detection, control by sprinklers or other installations, and firefighting. In some cases these may involve built-in devices but activation is required by the fire, and active measures are subject to malfunction, e.g. because of outside interference.

Buildings are usually classified according to intended use, occupancy, size and location; nevertheless, the fire regulations are aimed primarily at the safety of human occupants. If special circumstances apply and the owner requires additional standards, such as preservation of the structure, then these will be over and above the minimum regulatory requirements. In this respect there is currently a need for a more rational approach to the specification of fire safety in buildings, which takes account of risk, and aims at the optimum economic level of protection.

Since 90 per cent of all fatalities in fires occur as the result of smoke and toxic gases, it is particularly important for evacuation routes to be provided, with floors and staircases remaining free of smoke and properly illuminated. Smoke and fire doors should shut off floors and protect staircases. The distance from any room to a staircase should not be greater than about 30 m. Staircases must be designed to withstand a fire for 90 minutes. In all cases two evacuation routes must be provided. Up to a height of 20 m, fire brigade ladders may be considered to provide a second escape route.

Design models for pedestrian evacuation routes in high-rise buildings are treated in Ref. 6.3. In an example, the evacuation time for an administration building with 21 floors and three staircases is calculated at about 15 minutes for approximately 40 people per floor.

6.1.2   Losses and Costs

The national financial loss due to fire is made up of the costs of loss due to damage, and the costs of all investments in fire protection measures. The main components are as follows.

Losses:

- direct economic losses (structures and contents);
- indirect economic losses (effect on production, etc.);

- costs of administration of fire insurance companies;
- cost of death and injury.

Investments:

- cost of fire prevention measures;
- cost of alarm systems, extinguishing systems and other firefighting and safety devices;
- fire brigade costs.

The relative proportions of these components of national loss, as given in Table 6.1, indicate that substantial investments are made to reduce fire damage. It is clear that additional measures are justified only if the sum total of the losses is thereby reduced (Fig. 6.1). Although the data in Table 6.1 come from the late 1960s,[6.4] the relative values quoted should remain reasonably constant and comparable, in contrast with total figures which are affected by currency movements and inflation.

In limited applications, including the assessment of fire resistance requirements, building costs can be estimated, together with expected losses from damage or collapse, and used to produce economically optimal solutions. Such calculations must take account of the future economic life of the building and the fact that extra protection, e.g. in the form of sprinklers, can be added at a later date if it is not justified initially.

The major goal of fire safety, for governments as well as for builders and occupiers, is to maintain public confidence in the use and occupation of buildings. In certain buildings, there are additional economic and strategic reasons for ensuring high standards. For example, a large factory providing employment for the major part of a small community, a nuclear power plant and an important and sensitive submerged road tunnel clearly have special regional significance. Governments cannot accept the risk of a large-scale disaster affecting a whole community, and therefore have to legislate beyond what is strictly necessary to safeguard human life in a single building. Sometimes the specified objective may be to allow quick and simple repair, or avoid a repetition of a particular event.[6.5,6.6]

Table 6.1 Estimates of annual fire costs in various countries (per cent)

	Canada	The Netherlands	United Kingdom	United States
Losses				
Loss of property	30.3	26.3	30.0	29.5
Loss of use	–	12.3	–	–
Loss by injury	0.3	–	0.2	4.7
Expenditures				
Suppression	34.3	15.4	26.8	32.2
Prevention and protection	14.8	15.4	16.0	13.7
Insurance	20.2	30.8	26.8	19.5
Research and development	0.1	0.3	0.2	0.4

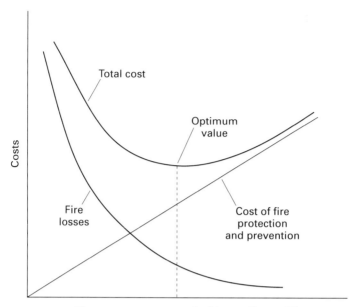

Measures to prevent and mitigate effects of fire

Fig. 6.1 Costs and losses

Fortunately the risk of fire is generally low. While it remains so, people are reluctant to suffer too much inconvenience. Systems that are installed solely for the purpose of fire safety should therefore interfere with normal usage as little as possible, yet harmonize with public feelings on acceptable risks.

On the other hand, the risks of fire are not always obvious. As an example of unexpected risk, a person sleeping in a room adjacent to a fire is highly endangered because the concentrations of CO and CO_2 can rapidly reach fatal levels, well before the increase in temperature would wake the sleeper.

A major contribution to both fire prevention and fire protection is the development of an awareness of fire hazard in those who are in a position to do something to reduce it, be they the designer, the architect, legal authorities or members of the public. Building requirements concentrate on fire protection, on the assumption that fires will occur and that their consequences have to be mitigated. This contrasts markedly with the concern of some authorities for identifying the 'cause' of a fire. Causes are particularly relevant in questions of liability and hazard analysis but are, in general, outside the scope of present building law, although heating and electrical systems and their maintenance are major exceptions.

6.1.3   Practical Conclusions

Scientific studies of fire risks have been undertaken over many years, and good progress has been made in translating the results into design practice, at least for the structural design for fire. However, progress has been slower for the broad field of fire safety. Apart from risk analyses performed by experts for unusual buildings such as nuclear power plants, fire risk assessment is treated as a regulatory issue. There are two main problems with this, as follows.

1. Experts are reluctant to translate the results of scientific investigations into simple design rules. The difficulty is that simple but economic rules may underestimate risk substantially in some cases because of the complexity of the physical phenomena.
2. Building authorities are hesitant to relax prescriptive regulations in favour of a proper engineering approach. This is because it is very difficult to translate prescriptive regulations into risk-oriented methods, in such a way that different, alternative solutions can in fact be verified as adequate. An example would be where two design teams use different solutions to achieve a comparable level of safety for the same building.

A highly courageous and complex investigation has been carried out in Australia to address the second problem.[6.2] Practical experience with this anticipated system will be of value worldwide.

6.2   Structural Fire Design

6.2.1   Fire Severity

The severity of a fire in a structure is dependent on many factors. Some are related to the pattern of the fire, while others depend on the effectiveness of measures available to control and extinguish the fire. However, building authorities assume for design purposes that a structure will be subjected to the worst-case fire, and indeed this approach is supported by the conventional wisdom of fire brigade experience, which is that any kind of combustible material will eventually burn.

The severity of a fire depends on three main factors:

*   available fuel;
*   ventilation, i.e. air supply available to promote its growth;
*   the characteristics of the compartment where the fire commences.

For a given fire load, the severity of the fire can vary by a factor of two or three depending on the ventilation and the configuration of the area. For practical purposes fire severity is expressed in the ISO 834 Specification[6.3] by means of a standardized temperature–time relationship which is also used in many national code regulations. Laboratory tests are conducted in furnaces using this relationship for the purpose of establishing the fire resistance of prototype structural elements.[6.7] This experimental approach has also been used to investigate the effects of various design variables, and much of the information available for design is derived from this type of study. Standard furnace tests are the main source of tabulated data in design codes on the fire resistance of different elements. Typical results from fire tests are shown in Fig. 6.2.

Current building controls in most countries express the standard of safety for a building in terms of the fire requirements for individual elements. The intention is

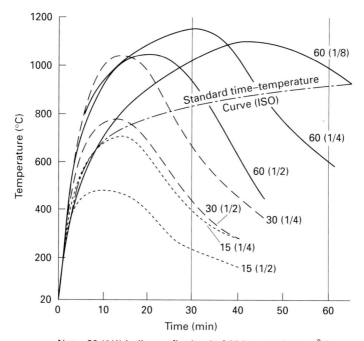

Note: 30 (1/4) indicates fire load of 30 kg wood per m² floor
area, window area a quarter of the wall in which windows are
situated.

Fig. 6.2 Average temperature development with wood cribs in fire tests

to ensure that the building structure does not collapse and that the separating
elements are maintained during a fire. Fire resistance requirements are expressed
for specified periods, ranging from 30 to 240 minutes. These periods do not
signify the duration of an actual fire. A 60 minute fire resistance does not thus
imply that a construction is expected to withstand a fire of 60 minutes duration,
but rather that it will withstand a fire of longer or shorter duration whose severity
corresponds to the 60 minute fire test.

   According to the limit states philosophy a strength limit state is reached when a
member can no longer carry its design load when combined with any additional
thermally-induced loading. Other serviceability limit states may also have to be
considered, such as those relating to the integrity of elements which provide
horizontal or vertical barriers to the passage of fire. The latter may, for example,
be required to prevent the progress of the fire through the gap between a floor slab
and the suspended ceiling, or through holes provided in walls or floors for service
pipes. A limit on deformation may also be used as a criterion for stability. In some
special cases where re-use of the structure is important, additional criteria may be
introduced which relate to repairability.

   The growth of a fire, illustrated in Fig. 6.3, can be divided into two phases,
called the pre- and post-flashover periods (Fig. 6.3(c)). There is a good chance
of being able to stop the fire in the pre-flashover period provided early

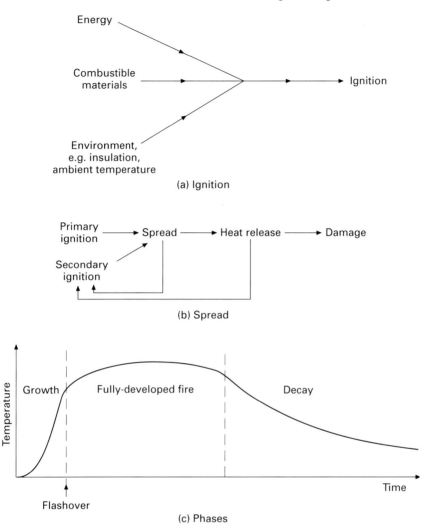

Fig. 6.3 Development of a fire

fire detection is followed by automatic alarm and firefighting with extinguishers or sprinklers.

The post-flashover period is characterized by a sudden rise in room temperature to above 550°C with the fire uniformly encompassing the room. It is a critical situation because the ability of the room or the fire compartment to contain a serious fire and prevent its spread depends on the fire endurance of the boundary components such as walls, doors and floor–ceiling assemblies. Good compartmentation is therefore crucial in minimizing fire risk. Of particular interest is the possibility of a fire spreading outside of the building and along the façade, and attacking external loadbearing columns. This is shown in Fig. 6.4.

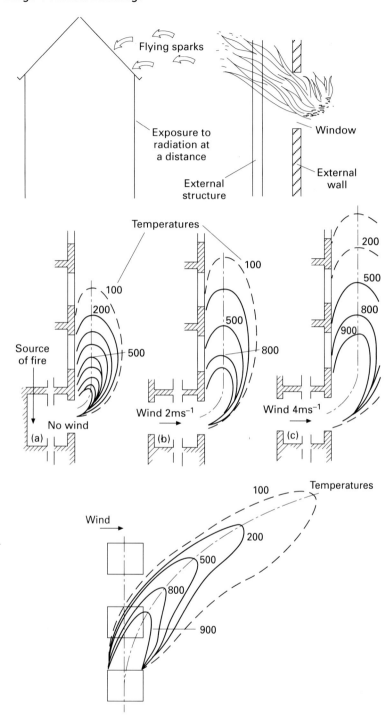

Fig. 6.4 Spread of fire through openings on outside of building

6.2.2   Structural Response

Building regulations in most countries provide for a minimum fire-protection strategy. Additional standards contain design aids for fire in the form of tables for dimensioning beams, slabs, columns, tension members and the like.[6.1] New standards such as the Eurocodes allow for additional computational methods[6.8] to predict the development of fires in buildings and the response of the structure to the fire.[6.3]

Structural requirements depend on the dimensions of the fire compartment, its use, and the individual functions of the various structural components. Structural requirements cover:

- the separating function;
- the loadbearing capacity of the structure.

After exposure to fire, a building may require demolition. If repairability is a design requirement, then limits on acceptable damage, residual deformation and degradation of material properties need to be specified.

In the design of a concrete structure for fire, an assessment of the loads and heat exposure is required, together with an analysis of structural response and an appropriate choice of structural system and components, including joints and supports. The fire exposure may be representative of a class of buildings and occupancy, or it may apply specifically to a particular project.

Material limit states for concrete and steel can be expressed in terms of a 'critical temperature', at which a specified deformation or rate of deformation is exceeded in a fire test specimen under constant stress. If the critical temperature is reached in a critical area of a loadbearing structural element, failure is initiated by compression, tension or bending.

Other failure modes also have to be considered, such as shear and bond, lack of rotation capacity over the interior supports of flexural systems, and explosive spalling of small-sized members.

In treating the separating function of floors and walls, limit states are specified for thermal insulation and structural integrity against fire penetration. It needs to be emphasized that structural integrity cannot be checked by calculation, but only by test.

Integrity implies the absence of cracks, orifices or openings through which flames and hot gases can pass, thus allowing the fire to break out on the other side. Retention of integrity is a design requirement for joints and junctions which should not open up as a result of excessive deformation of components. Occasionally, spalling of the concrete can lead to the formation of apertures in thin sections, with heat transfer through the construction. A fund of information on such phenomena is available in many countries and provides the basis for many national codes.

In a formal manner, three methods of assessment can be distinguished for use in structural fire design.

- *Assessment Method I.* This is based on the standard heating conditions as formulated in ENV 1992[6.1] or in ISO 834.[6.4] The design criterion is that the fire resistance time of a building component is equal to or greater than that required by the relevant regulation. For design, the dimensions of structural elements are obtained from tabulated data, taking into account the loads and the required fire resistance time.[6.1]

- *Assessment Method II.* This is based on the concept of an equivalent fire exposure time which is used to relate the effects of a non-standard compartment fire to those of the ISO 834 standard fire and the related 'standard heating conditions'. However, the design criterion is that the fire resistance of a structural component, determined either by experiment or calculation, is equal to or greater than the equivalent time of fire exposure. This equivalent time relates the non-standard compartment fire exposure to the standard heating conditions in such a way that the effect on the component is the same. Figure 6.5 gives further details.

- *Assessment Method III.* This method uses an analysis of the structure when subjected to a non-standard compartment fire exposure. Assessment of the temperature is based on heat and mass balance equations, and account is taken of factors such as the fire load characteristics, ventilation conditions and the thermal properties of the structural components surrounding the fire compartment.

National fire standards usually give basic information on the above three approaches. At the present time, evaluation according to methods I and II does not usually differ significantly; however, method III may need expert approval.

As a result of the analysis of fire tests, carried out in accordance with ISO 834 on a large number of structural components, tabulated data have been made available which simplify the assessment of the fire resistance of commonly used structural elements.[6.1] As a further design aid a simplified analysis can be made of the moment capacity of reinforced concrete members exposed to fire.

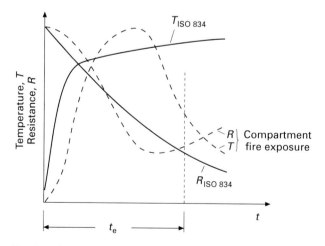

Fig. 6.5 Relation between 'compartment fire exposure' and 'standard heating conditions' (ISO 834), defining the 'equivalent time of fire exposure' t_e

6.3   Design and Detailing

6.3.1   Our Mission

Until recently, structural fire design was concerned purely with detailing, and more specifically with the choice of dimensions in accordance with tabulated data. A full analysis of the response of a structure to the development of a fire is now becoming possible, and this will be a major step towards the rational, economic and reliable design for fire. Nevertheless, the normal steps in structural design, namely

- concept design,
- detailed design calculations and
- detailing,

also need to be applied in the case of design for fire.

In normal design, detailing implies the translation of the design calculations into detailed drawings. This is only one aspect of detailing. Another aspect is that many influences (often minor) are allowed for in detailing which are not easy to analyse and quantify, but which are handled on the basis of experience, test data or 'how to build' knowledge which has emerged over the years.

The concept design and the detailing process together take care of those actions and situations which are unlikely to occur, and which are therefore not calculated in detail, but which require some consideration in design.

For this reason the authors consider that concept design and detailing are as important in fire design as the detailed design calculations. In this respect, designing and detailing for fire refer to the following:

- making the structure robust in relation to fire;
- designing for thermal expansions in the building;
- detailing for thermal expansions;
- compartmentation.

6.3.2   Robust Structures

What is a robust structure in the context of fire design? The concept of robustness is not yet well-defined even for normal structural design, and rather than deal with general concepts this discussion will emphasize several specific aspects.

Local Fire

Where fire resistance requirements do not strictly apply to the load bearing system, as in a one-storey building or a building with automatic sprinkler systems, the structure nevertheless should not collapse due to a small, local fire. This requirement would often be satisfied automatically because of the inherent fire resistance of normal concrete members which are not too slender. Without observing specific fire design requirements, concrete members are likely to have 30 to 60 minutes of fire resistance for the standard exposure situation.

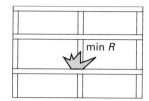

Providing for minimum fire resistance

Bending member transformed to tension member

Supported beam transformed to cantilever

Local collapse

Fig. 6.6 Performance in local fire

An alternative is to provide for a change in structural action, for example with flexural members acting in tension (Fig. 6.6). Other concepts may allow for local collapse, but ensure that the collapse region is limited.

Failure Mode

Although a structure may not survive a fully developed fire, gradual collapse, or collapse preceded by visual and acoustic warnings, is preferred in view of the safety of firefighters (Fig. 6.7). In this respect, normal concrete is preferable to aerated concrete. Stability failure may occur suddenly, and is undesirable in contrast with a strength failure preceded by large deformation.

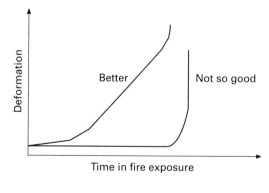

Fig. 6.7 Increase in deformation with time

Supports and Joints

Related to failure mode is the requirement that flexural failures should occur at in-span regions and not at supports or joints. Thus, members should not slip off supports, and the joints of prefabricated composite structures should be overdesigned:

$$R(\text{joint}) > R(\text{span})$$

Quality of Workmanship

Fire resistance should not be unduly affected by minor differences between the intended design and the on-site execution. For normal concrete members this is usually the case. However, problems can occur when lightweight frame-panels act as separating members or fire-protection screens (walls or suspended ceilings). Poor fixing of the panels and building service installations can easily ruin their separating and protecting functions.

6.3.3   Thermal Expansion

In principle, design for thermal expansion only requires rough estimates of how the structure will expand and deform under fire, although it may be necessary to consider also the cooling down period.

Thermal expansion and deformation depend mainly on the temperature increase in members during fire exposure. Figure 6.8 gives thermal gradients in concrete slabs under standard fire exposure,[6.1] which may also be considered as an equivalent time for fire exposure.

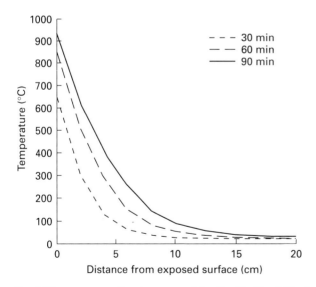

Fig. 6.8 Temperature gradient in concrete slabs after 30, 60 and 90 minutes of standard fire exposure

The thermal expansion of concrete can be expressed as a function of temperature approximately as follows:

$$\frac{\Delta l}{l} = 18 \times 10^{-6}\Theta_c \text{ for siliceous concrete, } \Theta_c \leq 700°C$$

$$\frac{\Delta l}{l} = 12 \times 10^{-6}\Theta_c \text{ for calcareous concrete, } \Theta_c \leq 800°C$$

$$\frac{\Delta l}{l} = 8 \times 10^{-6}\Theta_c \text{ for lightweight concrete}$$

As a comparison, siliceous concrete has values of about $10 \times 10^{-6}\Theta_c$ for Θ_c between 20 and 200°C.

For steel:

$$\frac{\Delta l}{l} = 14 \times 10^{-6}\Theta_s \text{ for reinforcing steel}$$

$$\frac{\Delta l}{l} = 12 \times 10^{-6}\Theta_s \text{ for prestressing steel}$$

For the simple assessment of normal reinforced concrete in fire, we may take:

$$\frac{\Delta l}{l} = 15 \times 10^{-6}\Theta_c$$

Expansion Joints

As a rough guide, the minimum width e of an expansion joint for fire should be

$$e = 50 \text{ mm, at 30 m spacing}$$

The need for expansion joints, and their spacing and widths, depend on:

- the expected severity of fire exposure;
- the proposed compartmentation;
- the structural design.

Obviously, if there is a sprinkler system on each floor or if fire loads are small, expansion joints may be dispensed with.

A slightly better guide is to provide expansion joints to allow for 50 per cent of the calculated thermal expansion.

Further issues regarding expansion joints include the following. In Fig. 6.9 the top storey of the building is separated into relatively small fire compartments, while the lower storey has just one large compartment. It is interesting to compare

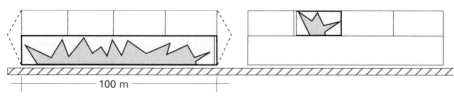

100 m

Fig. 6.9 Different situations concerning compartmentation

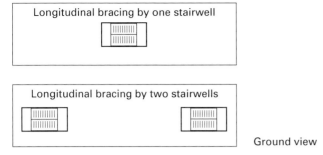

Ground view

Fig. 6.10 Effect of horizontal restraint

the effects of a fire in the lower storey and in a compartment in the upper storey. If the average increase in temperature in the slab is 200°C the potential expansion is in the order of 300 mm in the slab above the ground floor, but only about 80 mm for the roof slab. This ignores constraints and effects of curvature. Irrespective of the exact expansion which occurs, expansion joints are clearly more important in the slab above the ground floor than in the roof slab.

It is interesting to compare two different configurations for the horizontal constraints acting on the slab (Fig. 6.10). In one case there is a single stairwell in the middle of the building; in the other case there are two stairwells, one at either end of the building. The first arrangement clearly provides the minimum constraint to the slab and allows maximum expansion. The second case gives maximum constraint and minimum expansion. Disregarding the rest of the structure, no expansion joint would be required in the first case, whereas expansion joints would be needed in the second case to limit the compressive forces in the slab and the resultant forces on the stairwells. It should be noted that expansion joints in separating walls and slabs need special detailing.

Effects on Columns

If we now consider the columns supporting the slab in the case of the central constraint provided by the staircase, severe displacement at the column heads is likely to occur, as in Fig. 6.11, with horizontal forces acting on the columns. Expansion joints would be needed to protect the columns. On the other hand, in the design with stairwells at each end, the columns would be virtually unaffected, irrespective of the presence of expansion joints.

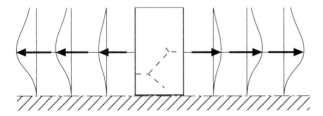

Fig. 6.11 Effect of central stairwell on surrounding columns (no expansion joint)

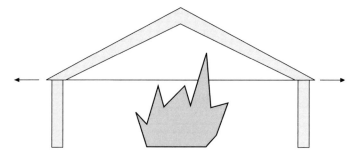

Fig. 6.12 Failure of tension member in fire

Tension Members

Tension members which are used for bracing or for supporting other members fail to fulfil their function when they are exposed to fire and expand. This is illustrated in Fig. 6.12.

Supports

When exposed to fire a simply supported beam expands and also deflects because it is heated on three sides (Fig. 6.13). After cooling down, the expansion may disappear, but the deflection remains. The beam may slip off a support which is not large enough.

Long Members

Design aids are usually derived from test results in which the size of specimen is limited by the dimensions of available test furnaces. Design data thus apply typically to moderate member dimensions such as occur in 'normal' buildings.

For buildings with small fire compartments, the effects of thermal expansion are small and are covered by the design rules. This is not necessarily the case for large structures with 'long' members. If the fire load in a large structure is small in comparison with the compartment size then there should not be significant problems. However, for high fire loads in large buildings, accurate fire design calculations should be performed to assess the thermal expansions, deformations and constraint forces.

As an example, consider a separating wall of 50 m height in a high bay warehouse. If the fire is not suppressed by automatic sprinklers then a severe fire

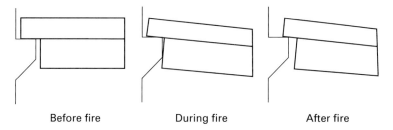

Before fire   During fire   After fire

Fig. 6.13 Beam support area before, during and after fire exposure

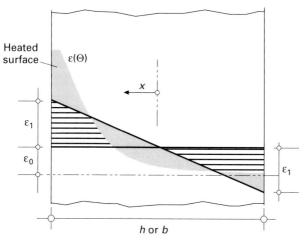

Fig. 6.14 Temperature-dependent strains

can produce the temperature gradients shown in Fig. 6.14 in the wall for a period of 90 minutes.

The resulting thermal expansion is not overly large, say 150 mm for an average temperature increase of 200°C. The temperature gradient produces internal stresses as well as curvature in the wall. Strictly, expansion and curvature should be calculated on the basis of equilibrium of the internal stresses:

$$\int \{\sigma[\varepsilon(\Theta)] - \sigma[\varepsilon_0]\} dx = 0$$

$$\int x \left\{ \sigma[\varepsilon(\Theta)] - \sigma\left[\varepsilon_1 \frac{2x}{b}\right] \right\} dx = 0$$

If a linear stress–strain relation is assumed and the dependency of the modulus of elasticity on temperature is ignored, then the resulting stress estimates are on the safe side. The error is larger for curvature than for expansion. If the non-loadbearing wall is 250 mm thick equilibrium occurs with strains of around $\varepsilon_1 = 0.004$. This results in a radius of curvature of only about 30 m when calculated from $1/R = 2\varepsilon_1/b$ (Fig. 6.15). A more precise calculation gives a value about twice this magnitude, but the result is irrelevant when the wall is 50 m high. This numerical exercise was chosen to be drastic; in practice such a wall would

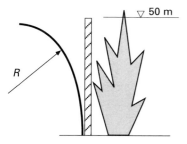

▽ 50 m

Fig. 6.15 Deformation of wall free to rotate

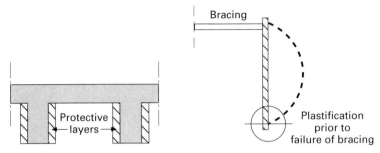

Fig. 6.16 Tentative guidance for design of high walls

usually be integrated with columns and of course the radius R increases with the width b of the wall.

Nevertheless, the full benefit of a large b value (equal to the column width b_c) is only achieved if the columns are on the cooler side of the wall, and not exposed to fire on three sides. What practical conclusions can be drawn for the design of high walls? Without systematic investigation, only the following tentative guidelines can be given (Fig. 6.16).

1. If the wall separates areas with high fire loads on one side and low fire loads on the other, then the columns should be located in the low fire-load area.
2. If this is not possible, protective layers can be applied to the column, so that full exposure only occurs on one side (this is not a typical 'concrete' solution).
3. The wall should be designed and braced in such a way that plastification would occur at the column foot prior to failure of the bracing.

Similar considerations apply generally to very long columns which are exposed to fire on one or three sides. In normal buildings the deformations induced in columns by asymmetric heating are compensated by lower temperatures as compared with the case of exposure on all four sides. With increasing column length the deformations increase, whereas the benefit from lower temperatures is unchanged.

Spalling

Explosive spalling can occur during the first 10–20 minutes of severe fire exposure. This is governed by the water content of the concrete, its porosity and the level of loading in conjunction with the restrained thermal expansions. Spalling may be avoided by limiting stresses and keeping the water content of the concrete to less than 4 per cent of the weight.

6.3.4   Compartmentation

Fire damage is mostly due to inadequate compartmentation, rather than to poor fire design of individual floors and walls, to achieve the separating function. Providing fire resistance for concrete members as such is usually not very expensive, but where compartmentation requires openings in walls and floors to

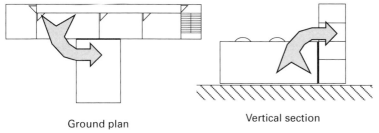

Ground plan   Vertical section

Fig. 6.17 Fire penetration from one part of the building to another part

be secured (doors, ducts, shutters) the cost may be considerable and the result may turn out to be useless because of elementary mistakes made on site and during use, as well as in design.

Mistakes in Design
Frequent mistakes are geometrically incomplete compartments in areas where the geometry of the structure is not straightforward, and in particular where the possibility of fire propagation across façades and roofs has not been adequately considered (Fig. 6.17). Serious mistakes can be encountered in the design of air-conditioning systems and building services in general. This is mostly due to lack of communication between the structural designer and the designers of the building services.

Mistakes on Site
The compartmentation envisaged by the designer is often not realized on site. The finish in areas which are easily accessible, and in particular at head height, is usually excellent with regard to separating members. On the other hand, areas which are difficult to access and thus control, such as walls extending above suspended ceilings (without any fire rating) and below cable floors (Fig. 6.18) are often in a deplorable state in regard to their separating function. Often site managers are not adequately informed about the required compartmentation.

Mistakes during Use
The compartmentation as originally planned is often obstructed during use. The best known example is the fire door which is forced to remain open to provide

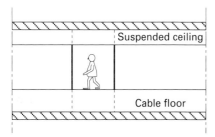

Fig. 6.18 Incomplete compartmentation above suspended ceilings and below cable floors

easy access. In buildings where electric or electronic cables are newly or repeatedly installed, it is common to forget to close the seals in compartment walls and floors. Maintenance of devices such as fire shutters is often not performed. Finally, in the course of time, the original compartmentation concept is simply forgotten, with the result that new installations such as ventilation ducts and transport systems are introduced without provision for the automatic barriers needed for fire.

Documentation

Some of these mistakes can be avoided by proper documentation of fire compartments on the general drawings, or by the use of specific fire-protection drawings. Such documentation serves several purposes:

1. It allows the designer to reduce the risk of geometrically inadequate compartments.
2. It provides information to the designers of building services on compartmentation.
3. It provides information to the site manager.

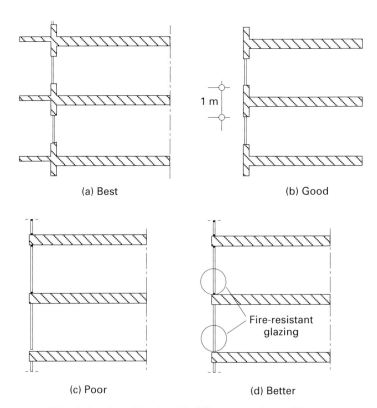

(a) Best   (b) Good

1 m

(c) Poor   (d) Better

Fire-resistant glazing

Fig. 6.19 Vertical section of façades with different separating qualities

4. It provides information to the client on the original compartmentation which needs to be maintained during use, or, in the case of alterations, to be replaced by an equivalent compartmentation.
5. It provides information to fire brigades and authorities.

Design of Façades

Windows in façades are always a problem with regard to the horizontal compartmentation of a building (Fig. 6.4). Breaching of fire compartments via windows is most effectively inhibited by fire-resistant parapets and balconies (Fig. 6.19(a)). Experience shows that traditional wall–window construction is also adequate, provided:

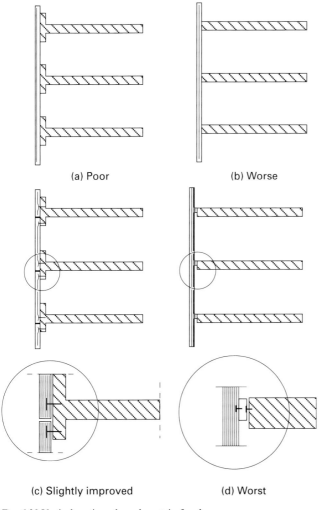

(a) Poor   (b) Worse

(c) Slightly improved   (d) Worst

Fig. 6.20 Vertical sections through curtain façades

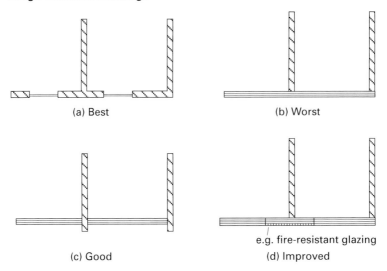

Fig. 6.21 Horizontal sections through face of building

- fire-resistant parapets are used;
- the flashover distance is at least 1 m;
- no combustible materials are used in the construction of the façade.

Nevertheless, with increasing building height, fire brigades are less likely to control fire propagation along the facade, so that flashover distances need to be increased for high-rise construction. Various treatments of the face of the building are considered in Figs 6.20 and 6.21.

6.4   Summary

In the development of an effective fire design procedure, the following items have to be incorporated:

- a risk-oriented approach to fire-protection measures, as outlined above in Section 6.1 and envisaged in Ref. 6.2;
- engineering methods for assessing expected fire severity as well as for predicting structural response for the purposes of structural design, as identified in Section 6.2 and in ENV1992;[6.1]
- procedures for concept design and detailing which deal with fire as a physical scenario rather than as a regulatory intervention, as tentatively outlined in Ref. 6.3.

References

6.1 ENV1992 *Eurocode 2, Design of Concrete Structures, Parts 1–2, Structural Fire Design*

6.2 Building Regulations Review Task Force (Australia) 1991 Microeconomic reform: fire regulations (explanatory document describing the development of effective regulations for fire safety)

6.3 Council on Tall Buildings and Urban Habitat 1992 *Fire Safety in Tall Buildings* McGraw Hill

6.4 *ISO 834, Structural Fire Design*

6.5 Lie T T 1972 *Fire and Building* Applied Science Publishers, London

6.6 Kordina K, Jeschar R *et al*. Brandversuche in Lehrte. Schriftenreihe Bau- und Wohnungsforschung, Bundesminister für Raumordnung, Bauwesen und Städtebau (Fire tests in Lehrte)

6.7 Bechthold R 1977 Zur thermischen Beanspruchung von Aussenstützen im Brandfall. Doctoral Dissertation, Technical University Braunschweig (Thermal loading of outside columns during fire)

6.8 Petterson O 1978 Assessment of fire severity by calculation. FIP/CEB Report on methods of assessment of the fire resistance of concrete structural members

7 Applications of High-Strength Concrete (HSC)

B. Vijaya Rangan

High-strength concretes (HSC) have been utilized in a number of buildings in Australia, North America, Japan and Europe. Most of this application has been in the columns. This chapter reviews the material properties of HSC. The behaviour of HSC columns as observed in tests is then described. Methods for the strength calculation of HSC columns are presented. Both stocky and slender columns are covered. The chapter includes detailed design steps for reinforced concrete columns as well as for tubular steel columns filled with concrete. These design steps are illustrated by examples. Suitable guidelines for the design of transverse reinforcement are also given.

7.1   Utilization of HSC

The definition of HSC has varied over the years. According to the ACI Committee 363,[7.1] concretes having compressive strengths of about 40 MPa or greater can be considered as high-strength materials. In a recent report by FIP-CEB Working Group on HSC,[7.2] all concretes in the range 60–130 MPa are defined as HSC. In this presentation, concretes with cylinder compressive strength of 50 MPa or more are defined as HSC.

Concretes with compressive strengths in the range 60–100 MPa have been used in numerous structures in North America, Europe, Japan and Australia.[7.1–7.4] Randall and Foot[7.5] have reported that the average strength of the concrete mix used in the Pacific First Center in Seattle was about 125 MPa. The usage depends on the availability of raw materials for the economical production of HSC. Other factors that require attention are the pumpability of concrete and the quality control both on- and off-site.

The largest application of HSC in buildings has been for columns. For instance, Table 7.1 lists the Australian buildings that utilized HSC. The economic advantages of HSC in columns of buildings are well documented. Smith and Rad[7.6] have studied the cost of column versus concrete strength for a five-storey and a 15-storey building. They have concluded that when compared to the columns made of 30 MPa concrete the reduction in the column-construction cost is of the order of 26 per cent for 55 MPa concrete and 42 per cent for 83 MPa concrete. ACI Committee 363[7.1] has pointed out that using HSC with a minimum of longitudinal steel (i.e. 1 per cent) is the most economical solution.

Comprehensive information on HSC is available in the literature.[7.1–7.4,7.7]

Table 7.1 Australian HSC projects[7.3]

Project	No. of levels	Concrete strength (MPa)	Time to strength (days)	Member
Perth				
SGIO Atrium	12	65	90	Columns
R & I Tower	48	65	90	Columns
St George's Square	20	65	90	Columns
Queen Victoria Bldg	42	60	28	Columns
Central Park	58	70	56	Columns
Melbourne				
The Rialto Project	60	60	56	Columns and core
Shell House	35	60	56	Columns and core
		70	56	Caisson
530 Collins Street	43	60	28	Columns and core
120 Collins Street	55	70	90	Columns and core
Melbourne Central	55	60	28	Columns
		65	28	Core
		70	60	Core
Telecom	52	70	56	Columns
		80	56	Columns
		60	56	Core
Collins Exchange		65	91	Core
		60	28	Columns
Palladium	15	70	56	Columns
		80	56	Columns
		60	56	Core
Bourke Place	55	60	28	Columns and core
Commonwealth Centre	42	75	91	Core
		65	91	Columns
Frankipile	NA	70	28	Piles
Brisbane				
Waterfront Place	39	65	28	Columns
Sydney				
135 King Street	30	60	56	Columns

7.2   Material Properties of HSC

7.2.1   Mix Proportions

The mix proportions of HSC vary depending on the availability of local materials. For instance, the mix proportions of HSC used in the columns of the Pacific First Center in Seattle are given in Table 7.2.[7.5] In Table 7.3, the mix proportions of normal weight HSC available in Perth, Australia are summarized.[7.8] Similar results are reported in the literature.[7.1–7.4,7.7]

General guidelines on proportioning of HSC are available.[7.1,7.2] It is, however, necessary that the designers should work closely with the local ready-mix plants in order to produce the optimum mix required for their project. A great deal of cooperation and team effort between the engineer, owner, contractor, and concrete producer are essential for the successful application of HSC.

Table 7.2 Mix proportions of HSC used in columns
of First Pacific Center, Seattle[7.5]

Compressive strength at 28 days	115 MPa
Modulus of elasticity at 28 days	56 000 MPa
Slump	250 mm
Cement (ASTM type II)	534 kg m^{-3}
Fly ash (Type F)	60 kg m^{-3}
Silica fume	40 kg m^{-3}
Pea gravel (10 mm)	1070 kg m^{-3}
Sand	625 kg m^{-3}
Water	130 kg m^{-3}
Plasticizer (Daratard 40)	2 kg m^{-3}
Superplasticizer (WRDA 19)	10 kg m^{-3}

Table 7.3 Mix proportions of normal weight HSC
available in Perth, Australia[7.8]

Compressive strength at 43 days	85 MPa
Maximum aggregate size	10 mm
Initial slump	70 mm
Cement (Type A)	450 kg m^{-3}
Silica fume	40 kg m^{-3}
10 mm aggregate	930 kg m^{-3}
7 mm aggregate	340 kg m^{-3}
Sand	510 kg m^{-3}
Water plus admixtures	160 kg m^{-3}

7.2.2   Behaviour in Compression

Strength Gain and Curing

Similarly to low-strength concrete, the compressive strength of HSC increases
with age. Typical results are given in Table 7.4.[7.4] The rate of increase in strength
depends on the mix proportions and the type of curing. Generally, HSCs have
high early-age strengths.[7.1] For instance, after 24 hours the strength of the mix
reported in Table 7.3 was approximately 45 per cent of the nominal 28-day value.
High early-age strengths can permit early stripping of formwork and therefore
increase the speed of construction.

Table 7.4 Ratio of compressive strength of HSC
at various ages to 28-day strength

Age	Strength to 28-day strength ratio
3 days	0.55
7 days	0.75
28 days	1.00
90 days	1.12
1 year	1.20
2 years	1.25

Test data[7.9,7.10] show that the effect of curing conditions has a greater influence on the strength of HSC than low-strength concrete. If HSC is allowed to dry out before completion of curing there is a greater reduction in strength compared to low-strength concrete. An adequate curing regime for at least seven days is therefore recommended.

Stress–Strain Relation
The stress–strain relation for HSC is significantly different to that of low-strength concrete. The main differences are[7.2]:

- a more linear stress–strain relationship up to 90 per cent of the maximum stress;
- a slightly higher strain at maximum stress;
- a steeper shape of the descending part of the curve.

It has been suggested that these differences are due to the improved bond between aggregate and cement paste which results in a reduction in the extent of microcracking at lower levels of loading of HSC.[7.2]

Various expressions have been proposed to describe the stress–strain relation of HSC. Of these, the one suggested recently by Collins *et al.*[7.11] shows considerable promise. Accordingly, the stress σ_c is related to the strain ε_c by the expression

$$\sigma_c = k_3 f_c' \frac{\varepsilon_c}{\varepsilon_c'} \frac{n}{n - 1 + (\varepsilon_c/\varepsilon_c')^{nk}} \qquad [7.1]$$

where f_c' is the cylinder compressive strength and k_3 is the reduction factor to relate f_c' to the *in situ* concrete strength. Based on test data, Collins *et al.*[7.11] have recommended that

$$k_3 = 0.6 + (10/f_c')$$
$$\leq 0.85 \qquad [7.2]$$

In Eq. 7.1,

$$n = 0.8 + (f_c'/17) \qquad [7.3]$$

$$k = 0.67 + (f_c'/62) \text{ when } \varepsilon_c/\varepsilon_c' > 1$$

$$= 1.0 \text{ when } \varepsilon_c/\varepsilon_c' \leq 1 \qquad [7.4]$$

$$\varepsilon_c' = \left(\frac{f_c'}{E_c}\right) \frac{n}{n - 1} \qquad [7.5]$$

and

$$E_c = 3320\sqrt{f_c'} + 6900 \qquad [7.6]$$

Note that in Eqs 7.1 to 7.6, f_c' is expressed in terms of MPa. Figure 7.1 illustrates the influence of concrete strength on shape of stress–strain relations as predicted by Eq. 7.1.

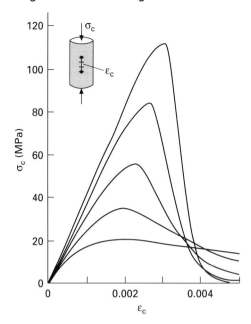

Fig. 7.1 Stress–strain relationship of concrete[7.11]

Modulus of Elasticity and Poisson's Ratio

Extensive data on modulus of elasticity of HSC are reported elsewhere.[7.1,7.2] Strictly, modulus of elasticity of HSC should relate to the stiffness of the aggregates in the mix. For normal-weight aggregates, measured values show good correlation with Eq. 7.6. For lightweight concretes, E_c may be obtained by multiplying the value given by Eq. 7.6 by the factor $(\rho/2400)^{1.5}$ where ρ is the density of concrete in kg m^{-3}. When compared to Eq. 7.6, the expressions given in various codes of practice overestimate the modulus of elasticity of HSC. In the absence of measured values Eq. 7.6 is therefore recommended.

Only limited data on Poisson's ratio of HSC have been reported.[7.1,7.2] These data indicate that a suitable value may be taken as 0.2. In the inelastic range, the relative increase in lateral strains is lower for HSC due to less microcracking.[7.2]

7.2.3   Tensile Strength

Customarily, the tensile strength of concrete is determined by performing the splitting tension test and the third-point flexural loading test. The tensile strength obtained from the split test is used in the calculation of shear strength of structural members. It is also related to the bond stress that exists in the concrete surrounding an embedded reinforcing bar. On the other hand, the flexure strength is utilized in the estimation of the cracking load of members.

The tensile strength test data usually show a large scatter. A number of equations have been proposed in the literature.[7.12] The following expressions given in the Australian Standard AS3600 – 1994 may be used to obtain lower-bound estimates:

$$\text{Splitting tensile strength} = 0.4\sqrt{f_c'}$$
$$\text{Flexural tensile strength} = 0.6\sqrt{f_c'}$$

where f_c' is expressed in terms of MPa.

7.2.4   Shrinkage

Limited information is available on the shrinkage of HSC. For a commercial mix available in Sydney, Australia, the results shown in Fig. 7.2 were obtained. The average shrinkage strain at 88 days was 450 microstrains. Based on these data, the 20-year final shrinkage strain was estimated as 750 microstrains.[7.13] This value is not significantly different from the final shrinkage strain of 710 microstrains calculated from the design data for low-strength concrete given in the Australian Standard AS3600. Other Australian data also indicates that the shrinkage of HSC is similar to that of low-strength concrete.[7.3]

In Fig. 7.3, the measured shrinkage strains of concretes used in 225 West Wacker Drive, Chicago as reported by Moreno[7.14] are given. These data show that shrinkage of HSC is in the range of 500–750 microstrains.

7.2.5   Creep

HSC creeps significantly less than low-strength concrete. For a commercial mix available in Sydney, Australia, the data given in Figure 7.4 were obtained.[7.13] The final creep coefficient, i.e. the ratio of creep strain to elastic strain for this concrete is estimated as 1.35, which is about 60 per cent of the value calculated using the design data given in the Australian Code AS3600 for low-strength concretes.

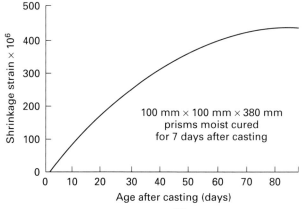

Fig. 7.2 Shrinkage of a commercial HSC[7.13]

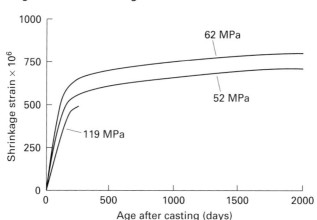

Fig. 7.3 Measured shrinkage of HSC[7.14]

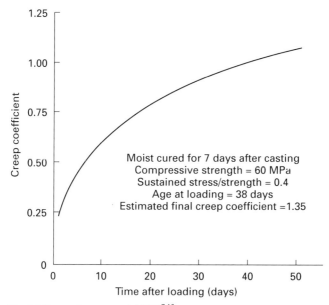

Fig. 7.4 Creep of a commercial HSC[7.13]

The measured creep strain of concretes used on 225 West Wacker Drive, Chicago as reported by Moreno[7.14] are given in Fig. 7.5. These data show that creep decreases with the increase of concrete strength.

7.3   HSC Columns

7.3.1   Columns under Concentric Compression

The general behaviour of HSC columns under concentric compression may be schematically represented as shown in Fig. 7.6.[7.15] In Fig. 7.6, the ascending

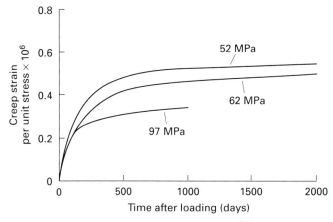

Fig. 7.5 Measured creep strain of HSC per psi of stress[7.14] (note: 1 MPa = 145 psi)

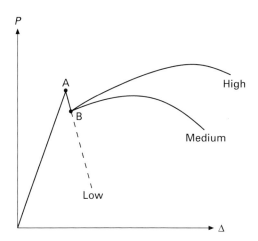

Fig. 7.6 Influence of low, medium and high amounts of transverse reinforcement on the behaviour of HSC columns subject to concentric compression[7.15]

branch of the load–deformation curve is a straight line up to the first peak load, A. The cover concrete usually spalls off at point A. After spalling of the cover concrete, there is a sudden drop in the load carrying capacity as indicated by the point B.

The descending branch of the load–deformation curve is significantly influenced by the quantity of transverse reinforcement in the column. When the volumetric ratio of the transverse reinforcement is small, no ductility is present. Test data show that HSC columns require large quantities of transverse reinforcement to achieve ductile behaviour. In Fig. 7.7 taken from Ref. 7.15, the ductility is represented by the ratio $\varepsilon_{85}/\varepsilon_{01}$, where ε_{85} is the axial strain in core concrete when the load drops to 85 per cent of the first peak load and ε_{01} is the axial strain corresponding to the first peak load. It can be seen that columns of different concrete strengths having the same transverse reinforcement index $p_s f_{yt}/f_c'$ possess almost the same axial ductility. Another parameter that influences the ductility is the distribution of longitudinal bars within the column cross-

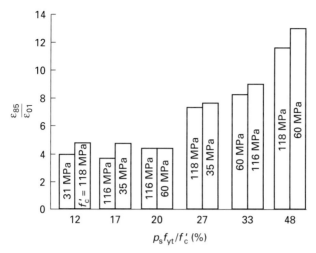

Fig. 7.7 Influence of transverse reinforcement index on the ductility of HSC columns subject to concentric compression[7.15]

section. For instance, Yong *et al.*[7.16] have observed that the ductility of tied HSC columns is improved when the cross-section contained not less than eight longitudinal bars.

The axial load capacity P_0 under concentric compression is usually given by

$$P_0 = 0.85f'_c(A_g - A_s) + f_y A_s \qquad [7.7]$$

The strength of HSC columns predicted by Eq. 7.7 seems to vary with the transverse reinforcement index $p_s f_{yt}/f'_c$.[7.15] For low values of $p_s f_{yt}/f'_c$, the predictions by Eq. 7.7 may be slightly unconservative. However, well confined columns may reach strength in excess of that predicted by Eq. 7.7. Collins *et al.*[7.11] have suggested that the factor 0.85 in Eq. 7.7 may be replaced by k_3 given by Eq. 7.2. This applies especially in the case of HSC columns with low levels of confinement. However, extensive test data reported recently by Cusson and Paultre[7.17] show that Eq. 7.7 may be acceptable provided that the column cross-section contains at least eight evenly distributed longitudinal bars.

7.3.2   Short Columns under Combined Axial Compression and Bending Moment

A number of investigators have reported on the behaviour of HSC columns subjected to combined axial compression and bending moment.[7.15,7.18–7.22]

The strength of a column section is customarily calculated by assuming a linear strain distribution over the depth of the section and considering the equilibrium of forces and moment.[7.23] In order to apply this procedure to HSC columns, two factors require attention. First, the ultimate compressive strain ε_{cu} at which extreme compression face of the column reaches failure should be known.

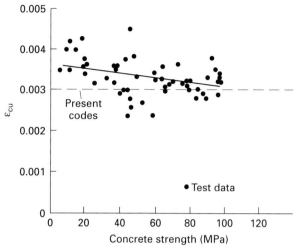

Fig. 7.8 Variation of ultimate compressive strain with concrete compressive strength[7.24]

Secondly, the actual distribution of compressive stresses in the concrete should be defined.

Figure 7.8 shows the variation of ε_{cu} with concrete strengths as reported in Ref. 7.1 based on test data obtained from experiments conducted on singly reinforced concrete beams or eccentrically loaded columns without lateral confinement steel.[7.24] It can be seen that $\varepsilon_{cu} = 0.003$, which is specified by the codes and standards,[7.25,7.26] represents satisfactorily the test results. Although this value is somewhat less conservative for HSC than for low-strength concrete, the scatter of test data shown in Fig. 7.8 does not justify a variation of ε_{cu} with the concrete compressive strength.

The distribution of compressive stresses in the concrete may be calculated by Eq. 7.1. Based on HSC-column tests, a modified expression for k_3 has been proposed:[7.22]

$$k_3 = 0.85 - 0.008(f_c' - 50) \qquad [7.8]$$

within the limits $0.65 \leq k_3 \leq 0.85$. In codes and standards,[7.25,7.26] the actual distribution of compressive stresses in the concrete is replaced by the equivalent rectangular stress block concept. The stress block has a uniform stress of $0.85f_c'$ and a depth equal to γ times the depth of neutral axis, where γ is given by

$$\gamma = 0.85 - 0.008(f_c' - 30) \qquad [7.9]$$

within the limits $0.65 \leq \gamma \leq 0.85$.

In recent years, a number of proposals have emerged to modify the stress block for HSC.[7.15,7.18,7.19] It is generally accepted that the uniform stress should be smaller than $0.85f_c'$ for HSC. The proposal by Li *et al.*[7.18] which may be included in the proposed revision to the New Zealand Standard,[7.27] is recommended. In this proposal, the depth of the equivalent rectangular stress block is given by Eq. 7.9 and the magnitude of the uniform stress is taken as $\alpha f_c'$ where

$$\alpha = 0.85 - 0.004(f_c' - 55) \qquad [7.10]$$

within the limits $0.75 \leq \alpha \leq 0.85$. Note that $\alpha = 0.85$ when $f'_c \leq 55$ MPa and $\alpha = 0.75$ when $f'_c \geq 80$ MPa. The reduction in the magnitude of the uniform stress to values less than $0.85f'_c$ for HSC is also recommended by the Norwegian concrete design code.[7.28]

No study has been made on the behaviour and the strength of HSC short columns subjected to biaxial bending and axial compression. For low-strength concrete columns, the method proposed in the Australian Standard[7.26] shows good agreement with the test results.[7.29] This method may be used for HSC columns provided that the equivalent rectangular stress block as defined by Eqs 7.9 and 7.10 is used.

7.3.3   Slender Columns

The strength of a slender column is affected by many factors such as column length, end restraint conditions, distribution of bending moment, creep of concrete and bracing condition of the column. In design, the strength of the slender column is usually calculated on the basis of an equivalent standard pin-ended column model. Therefore, studies on slender columns have concentrated on the behaviour and strength of this standard column.

The axial load capacity of a standard pin-ended slender column shown in Fig. 7.9 can be calculated using a stability analysis.[7.23,7.30] In order to perform such an analysis, a family of moment–curvature curves is required for a given value of the axial thrust. If the deflected shape of the column is assumed to be a particular mathematical function, then the curvature κ at mid-height is related in a simple manner to the mid-height deflection Δ.

Using a sinc function for the deflected shape $v(x)$,

$$v(x) = \Delta \sin \left(\frac{\pi x}{L_e} \right) \qquad [7.11]$$

The curvature κ at mid-height where $x = L_e/2$ is given by

$$\kappa = \frac{\pi^2}{L_e^2} \Delta \qquad [7.12]$$

where Δ is the deflection at mid-height.

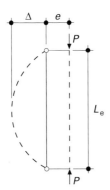

Fig. 7.9 Standard pin-ended slender column under short-term load

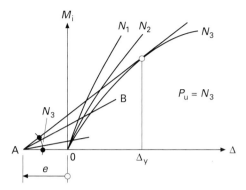

Fig. 7.10 Family of moment–deflection curves for stability analysis of slender columns

For a given column and for a chosen value of the axial thrust, the moment–curvature diagram is first converted to the moment–deflection diagram by means of Eq. 7.12. A sequence of increasing values of axial thrust, P, is then chosen and the corresponding family of curves of internal moment M_i versus deflection Δ is constructed as in Fig. 7.10.

The relationship between Δ and the external moment M_e is as follows:

$$M_e = P(e + \Delta) \qquad [7.13]$$

where e is the eccentricity of the axial thrust.

Equation 7.13 is represented in Fig. 7.10 by the line AB at a slope numerically equal to P, for any one of the chosen values of $P = N_1, N_2, N_3$ etc. The distance AO is numerically equal to the eccentricity e.

The point of intersection of the line representing Eq. 7.13 and the moment–deflection curve, that is, the intersection of lines of internal and external moment, represents the state of equilibrium and gives the value of deflection Δ for this load level. By constructing rays through point A at slopes which correspond to the sequence of increasing values chosen for P, a sequence of equilibrium points can be found. These equilibrium points can then be used to construct the load–deflection (P–Δ) curve shown in Fig. 7.11.

The axial load capacity of the column P_u is then given by the peak load, as shown in Fig. 7.11. The corresponding deflection is Δ_y.

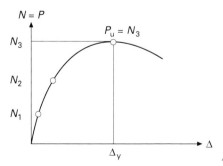

Fig. 7.11 Load–deflection curve

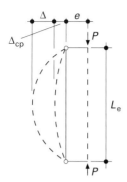

Fig. 7.12 Standard pin-ended slender column including creep deflection

In practice, a significant part of the design load applied on a column is sustained for a long period of time. The sustained load produces creep deflection, Δ_{cp} (Fig. 7.12) which causes a loss in load-carrying capacity of the column.

In the stability analysis, the creep deflection, Δ_{cp} is treated as an additional eccentricity. In order to calculate Δ_{cp} the method proposed elsewhere[7.31] to estimate the deflection of slender columns under sustained load may be used. The expressions recommended in Ref. 7.31 are as follows:

$$\Delta_{tot} = \frac{e}{[(P_c/P_\phi) - 1]} \tag{7.14}$$

where

$$P_c = \pi^2 \frac{EI}{L_e^2} \tag{7.15}$$

$$EI = \frac{\lambda E_c I_g}{1 + 0.8\phi_{cc}} \tag{7.16}$$

and

$$\lambda = \left[0.6 + \left(\frac{e_b}{8e}\right)\right] \leqslant 1.0 \tag{7.17}$$

In these expressions P_ϕ is the axial thrust due to sustained loads, ϕ_{cc} is the creep coefficient, e_b is the value of e corresponding to balanced failure in combined axial compression and bending, and Δ_{tot} is the final deflection at mid-height of a standard pin-ended column due to sustained load P_ϕ.

The total deflection is taken as the sum of the elastic component, Δ_e and the creep deflection, Δ_{cp}. From Eq. 7.14, the elastic deflection can be written as

$$\Delta_e = \frac{e}{[(P_{c0}/P_\phi) - 1]} \tag{7.18}$$

where

$$P_{c0} = \frac{\lambda \pi^2 E_c I_g}{L_e^2} \tag{7.19}$$

That is, Δ_e is the particular value of Δ_{tot} when $\phi_{cc} = 0$. Therefore, the creep deflection becomes

$$\Delta_{cp} = \Delta_{tot} - \Delta_e \qquad [7.20]$$

If we include the effect of creep deflection, the external moment M_e given by Eq. 7.13 is modified as

$$M_e = P(e + \Delta + \Delta_{cp}) \qquad [7.21]$$

The stability analysis described above has shown good correlation with test values obtained from 90 columns made of low-strength concrete[7.30] and 42 HSC columns.[7.21,7.22] In the case of HSC columns, the stress–strain relation for concrete in compression has been taken as given by Eq. 7.1, where the factor k_3 is calculated by Eq. 7.8.

For design purposes, the stability analysis has been simplified elsewhere.[7.32] Accordingly, if P_u is the factored axial load at an equivalent eccentricity e then the co-existing magnified factored moment M_e is given by

$$M_e = P_u(e + \Delta_y + \Delta_{cp}) \qquad [7.22]$$

In Eq. 7.22, the deflection Δ_y at failure has been approximated[7.32] by the following.

- For $P_u \geq \phi P_b$,

$$\Delta_y = \Delta_{yb}(\phi P_0 - P_u)/(\phi P_0 - \phi P_b) \qquad [7.23]$$

- For $P_u \leq \phi P_b$,

$$\Delta_y = \Delta_{y0} + (\Delta_{yb} - \Delta_{y0})(P_u/\phi P_b) \qquad [7.24]$$

where

$$\Delta_{yb} = (0.003 + \varepsilon_y)L_e^2/\pi^2 d \qquad [7.25]$$

$$\Delta_{y0} = 1.6\varepsilon_y L_e^2/\pi^2 d \qquad [7.26]$$

P_b is the particular axial load strength at balanced failure conditions, d is the depth of extreme layer of tensile steel measured from the compression face and ϕ is the strength reduction factor.

Based on the preceding information, the following steps are proposed for the design of HSC slender columns in braced frames.

1. Select a trial cross-section for the column. Calculate the effective length L_e of the column using the methods given in the codes.[7.25,7.26]
2. Calculate the eccentricity e for the equivalent standard pin-ended column from

$$e = k_m M_2/P_u \qquad [7.27]$$

where M_2 is the value of the larger factored end moment, k_m is given by

$$k_m = (0.6 - 0.4M_1/M_2)$$
$$\geq 0.4 \qquad\qquad [7.28]$$

and M_1 is the smaller factored end moment. The ratio M_1/M_2 is less than or equal to unity and is taken as negative when the column is in single curvature and positive for double curvature.[7.25,7.26]

3. Calculate the design strength interaction diagram for the column cross-section using the equivalent rectangular stress block as defined by Eqs 7.9 and 7.10.

4. Calculate Δ_{cp} by Eqs 7.14, 7.18 and 7.20 and Δ_y by Eq. 7.23 or 7.24. For these values of e, Δ_y, and Δ_{cp}, and given value of P_u, calculate M_e by Eq. 7.22. Check whether the design strength of the column cross-section is adequate to resist the combined effect of the factored actions P_u and M_e.

The preceding design method is illustrated by an example in Section 7.4.

7.3.4   Transverse Reinforcement Requirement

A number of proposals for the design of transverse reinforcement in HSC columns have been reported. Based on the research at the University of Canterbury[7.18], the following requirements are given in the proposed New Zealand Standard[7.27] for the design of transverse reinforcement in potential plastic hinge regions of HSC columns to confine the core concrete and to prevent premature buckling of longitudinal reinforcement.

When spirals are used, the volumetric ratio p_s should not be less than that given by the following equations.

- To confine the concrete,

$$p_s = \left(\frac{1.3 - p_t m}{2.4}\right) \frac{A_g}{A_c} \frac{f_c'}{f_{yt}} \frac{P_u}{\phi f_c' A_g} - 0.0084 \qquad\qquad [7.29]$$

- To prevent buckling of compressed longitudinal bars,

$$p_s = \frac{A_s}{2400 d''} \frac{f_y}{f_{yt}} \qquad\qquad [7.30]$$

In Eqs 7.29 and 7.30, $p_s = A_s/A_g$, A_s is the total area of longitudinal reinforcement, $m = f_y/0.85 f_c'$, A_g is the gross area of column cross-section, A_c is the area of concrete core of section measured to outside of peripheral transverse steel, d'' is the diameter of concrete core of section measured to outside of peripheral transverse steel, f_y is the yield strength of longitudinal reinforcement, f_{yt} is the yield strength of transverse reinforcement, f_c' is the concrete compressive cylinder strength, P_u is the axial load on column and ϕ is the strength reduction factor, equal to 0.85 if plastic hinging can occur or 1.0 if the column is protected from plastic hinging.

When rectangular hoops with or without cross-ties are used, the total area of transverse bars A_{sh} in each of the transverse directions within spacing s_h should not be less than given by Eqs 7.31 and 7.32.

- To confine the concrete,

$$\frac{A_{sh}}{s_h h''} = \left(\frac{1.3 - p_t m}{3.3}\right) \frac{A_g}{A_c} \frac{f'_c}{f_{yt}} \frac{P_u}{\phi f'_c A_g} - 0.006 \qquad [7.31]$$

- To prevent buckling of compressed longitudinal bars,

$$A_{sh} = \sum A_{te} \qquad [7.32]$$

In Eqs 7.31 and 7.32, h'' is the dimension of concrete core of the section measured perpendicular to the direction of the hoop bars to the outside of the perimeter hoop, s_h is the centre-to-centre spacing of hoop sets and $\sum A_{te}$ is the sum of areas of legs required to tie the longitudinal bars. The area of tie leg A_{te} required to tie the longitudinal bars reliant on it is

$$A_{te} = \frac{1}{16} \frac{\sum A_b f_y}{f_{yt}} \frac{s_h}{100} \qquad [7.33]$$

where $\sum A_b$ is the sum of areas of longitudinal bars reliant on the tie, f_y is the yield strength of longitudinal reinforcement, f_{yt} is the yield strength of transverse reinforcement and s_h is the centre-to-centre spacing of ties.

In Eqs 7.29, 7.30 and 7.31, A_g/A_c should not be taken greater than 1.2, $p_t m$ should not be taken greater than 0.4 and f_{yt} should not be taken larger than 800 MPa.

In the above, the centre-to-centre spacing of spirals should not exceed the smaller of one-quarter of the least lateral dimension of the column cross-section or six longitudinal bar diameters. For rectangular hoops, the centre-to-centre spacing should not exceed the larger of one-quarter of the lateral dimension of the column cross-section or 200 mm.

7.4   Design of HSC Columns – Example

The cross-section of a reinforced rectangular HSC column has the following details: $b = 400$ mm, $D = 600$ mm, $f'_c = 80$ MPa, $f_y = 400$ MPa, $A_{st} = A_{sc} = 3200$ mm^2 (i.e. four Y32 bars, area of each bar 800 mm^2), $d = 530$ mm, $d_{sc} = 70$ mm, equivalent eccentricity $e = 90$ mm, $P_u = 4500$ kN, $P_\phi = 2700$ kN, equivalent length $L_e = 10$ m, $E_c = 36\,500$ MPa, $E_s = 200 \times 10^3$ MPa, $\varepsilon_y = 0.002$, $\phi_{cc} = 1.5$ and $\phi = 0.6$.

Check the adequacy of the cross-section to carry the design loads.

Eq. 7.7: $P_0 = 0.85 \times 80(400 \times 600 - 6400) + 400 \times 6400$

$= 18\,445$ kN

$\phi P_0 = 0.6 \times 18\,445 = 11\,067$ kN

Eq. 7.10: $\alpha = 0.85 - 0.004(80 - 55) = 0.75$

Eq. 7.9: $\gamma = 0.85 - 0.008(80 - 30) = 0.45 < 0.65$, take $\gamma = 0.65$

Balanced Failure

With reference to Fig. 7.13, for balanced failure $\varepsilon_{st} = \varepsilon_y = 0.002$. From strain distribution,

$$d_n = \frac{0.003}{0.003 + 0.002} \times 530 = 318 \text{ mm}$$

For this value of d_n compression steel will yield.

Equilibrium of forces gives

$$P_b = (0.75 \times 80 \times 0.65 \times 318 \times 400) + (3200 \times 400)$$
$$- (3200 \times 400) = 4961 \text{ kN}$$
$$\phi P_b = 0.6 \times 4961 = 2977 \text{ kN}$$

By summing moments of forces about the plastic centroid,

$$M_b = 3200 \times 400(300 - 70)$$
$$+ 0.75 \times 80 \times 0.65 \times 318 \times 400 \left(300 - \frac{0.65 \times 318}{2} \right)$$
$$+ 3200 \times 400(530 - 300) = 1564 \text{ kN m}$$
$$\phi M_b = 0.6 \times 1564 = 938 \text{ kN m}$$
$$e_b = M_b/P_b = 1564/4961 = 315 \text{ mm}$$

Magnified Moment M_e

Eq. 7.25: $\Delta_{yb} = \dfrac{(0.003 + 0.002)(10 \times 10^3)^2}{\pi^2 \times 530} = 96 \text{ mm}$

Since $P_u > \phi P_b$,

Eq. 7.23: $\Delta_y = \dfrac{96(11\,067 - 4500)}{11\,067 - 2977} = 78 \text{ mm}$

Eq. 7.17: $\lambda = 0.6 + [315/(8 \times 90)]$
$= 1.04 > 1.0$, take $\lambda = 1.0$

Eq. 7.16: $EI = \dfrac{1.0 \times 36\,500 \times \frac{1}{12} \times 400 \times 600^3}{1 + 0.8 \times 1.5}$
$= 119\,455 \times 10^9 \text{ N mm}^2$

Eq. 7.15: $P_c = \dfrac{\pi^2 \times 119\,455 \times 10^9}{(10\,000)^2} = 11\,778 \text{ kN}$

Eq. 7.14: $\Delta_{tot} = \dfrac{90}{(11\,778/2700) - 1} = 27 \text{ mm}$

Eq. 7.19: $P_{co} = \dfrac{1.0 \times \pi^2 \times 36\,500 \times \frac{1}{12} \times 400 \times 600^3}{(10\,000)^2} = 25\,912 \text{ kN}$

Eq. 7.18: $\Delta_e = \dfrac{90}{(25\,912/2700) - 1} = 10$ mm

Eq. 7.20: $\Delta_{cp} = 27 - 10 = 17$ mm

Eq. 7.22: $M_e = 4500(90 + 78 + 17) = 833$ kN m

Adequacy of Cross-Section

It is necessary to check whether the cross-section is adequate to carry the combined effect of axial thrust $P_u = 4500$ kN and magnified moment $M_e = 833$ kN m. Note that $M_e/P_u = 833/4500 = 185$ mm $< e_b$; therefore the type of failure will be primary compression and the neutral axis depth d_n will be greater than the value calculated for balanced failure.

Another point on the strength interaction diagram is calculated when $d_n = d = 530$ mm. For this value of d_n, from the strain diagram (Fig. 7.13), $\varepsilon_{sc} > \varepsilon_y$ and $\varepsilon_{st} = 0$.

From equilibrium of forces,

$$P_n = (0.75 \times 80 \times 0.65 \times 530 \times 400) + (3200 \times 400) - 0$$
$$= 9548 \text{ kN}$$
$$\phi P_n = 0.6 \times 9548 = 5729 \text{ kN}$$

By summing moments of forces about the plastic centroid,

$$M_n = (0.75 \times 80 \times 0.65 \times 530 \times 400)\left(300 - \frac{0.65 \times 530}{2}\right)$$
$$+ (3200 \times 400)(300 - 70) = 1351 \text{ kN m}$$
$$\phi M_n = 0.6 \times 1351 = 810 \text{ kN m}$$

The strength interaction diagram between this point ($\phi M_n = 810$, $\phi P_n = 5729$) and the balanced failure point ($\phi M_b = 938$, $\phi P_b = 2977$) may be approximated as a straight line.

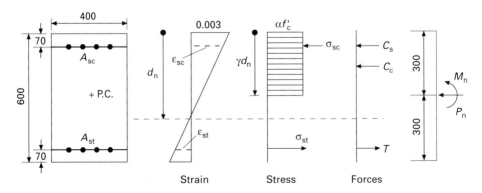

Fig. 7.13 Example: HSC column

The equation of this straight line is given by

$$\frac{\phi M_n - 810}{938 - 810} = \frac{\phi P_n - 5729}{2977 - 5729}$$

or

$$\phi M_n = 810 + \left(\frac{5729 - \phi P_n}{21.5}\right)$$

For the example column, if we substitute $\phi P_n = P_u = 4500$ kN in the above expression $\phi M_n = 867$ kN m > 833 kN m $(= M_e)$. Therefore, the column cross-section is adequate to carry the design loads.

7.5   Tubular Steel Columns Filled with HSC

Tubular steel columns filled with HSC are an attractive economic alternative to conventional reinforced concrete columns. Such composite columns offer a number of advantages during construction. These include simultaneous multilevel construction, reduced on-site labour and crane usage as a result of the lack of formwork required and the ease of assembly, and ease of concrete placement by pumping upwards from the base of the column resulting in no need to vibrate the concrete. All these advantages lead to reduced construction time and costs when compared to other forms of column construction.

Tubular steel columns filled with HSC behave in a ductile manner as the concrete is ideally confined. In Australia, a number of buildings have utilized this type of column.[7.33–7.35]

The strength interaction diagram of a tubular column filled with HSC and subject to axial thrust and bending moment can be calculated by performing the usual flexural analysis of the column cross-section. For this purpose, the composite column may be idealized as shown in Fig. 7.14.[7.36] Accordingly, the steel tube is split into two strips about the neutral axis. The area of steel in the compression zone, denoted by A_{sc}, is assumed to be lumped at its centroid. Similarly, the area of steel in the tension zone is A_{st} and the area of concrete in the compression zone is A_c. Both these areas are also assumed to be concentrated at their respective centroids. The concrete in the tension zone is neglected.

From Fig. 7.14, for equilibrium

$$P_n = C_s + C_c - T \qquad [7.34]$$

$$M_n = C_s z_{sc} + C_c z_c + T z_{st} \qquad [7.35]$$

where z_{sc}, z_c and z_{st} are the lever arms of forces measured from the plastic centroid, and $C_s = A_{sc}\sigma_{sc}$, $C_c = A_c\sigma_c$ and $T = A_{st}\sigma_{st}$. The steel stresses are related to the steel strains in terms of the depth of neutral axis d_n by

$$\sigma_{sc} = E_s(\varepsilon_1 d_{sc}/d_n) \le f_y \qquad [7.36]$$

$$\sigma_{st} = E_s\varepsilon_1 d_{st}/d_n \le f_y \qquad [7.37]$$

where E_s is the elastic modulus of steel, ε_1 and ε_2 are the strains at the extreme fibres (see Fig. 7.14), d_{sc} and d_{st} are the distances of the respective steel areas measured from the neutral axis and f_y is the yield stress of steel.

Since the concrete inside the tubular steel column is confined, Eq. 7.1 may not be appropriate to describe its stress–strain behaviour. Assuming that the σ–ε

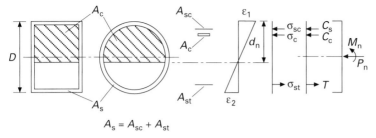

$$A_s = A_{sc} + A_{st}$$

Fig. 7.14 Idealization of tubular steel columns filled with concrete

relation for concrete is defined by Hognestad's parabola, the concrete stress σ_c is given by

$$\sigma_c = 0.85 f'_c [2(\varepsilon_c/\varepsilon_0) - (\varepsilon_c/\varepsilon_0)^2] \qquad [7.38]$$

where f'_c is the cylinder compressive strength, ε_0 is taken as 0.002 and ε_c is related to the depth of neutral axis by

$$\varepsilon_c = \varepsilon_1 (d_c/d_n) \qquad [7.39]$$

in which d_c is the distance of concrete area from the neutral axis.

The curvature κ is given by

$$\kappa = \varepsilon_1/d_n$$

or $\qquad \kappa = \varepsilon_2/(D - d_n) \qquad [7.40]$

Equations 7.34 to 7.40 can be used to calculate the moment–thrust–curvature relations for the composite column section.

When the column becomes slender, the strength can be calculated by performing the stability analysis as described in Section 7.3 (see Figs 7.10 and 7.11). Since the composite columns are ductile, the moment–deflection relation at the critical section may be approximated by elastic–plastic behaviour as shown by line OYY' in Fig. 7.15. The maximum load P_u at which equilibrium is possible is represented by the slope of the line AYB, which radiates from point A and becomes tangential to the M–Δ curve (Fig. 7.15).

For a standard pin-ended column (Fig. 7.12), the critical section is at mid-height of the column. The axial load capacity of the column P_u is, as before, related to the magnified moment M_e by Eq. 7.22.

In Eq. 7.22, Δ_y is computed by Eqs 7.40 and 7.12, and is governed by either ε_1 or ε_2 (Fig. 7.14) developing the yield strain of steel tube ε_y. The creep deflection Δ_{cp} is calculated by Eqs 7.14, 7.18 and 7.20. In these calculations the effective bending stiffness $\lambda E_c I_g$ may be taken equal to the bending stiffness $E_c I_{gt}$ of the transformed uncracked concrete section of the column at the mid-height.

Based on the preceding method, the following steps are proposed for the design of slender tubular steel columns filled with concrete.

1. Select a trial cross-section for the composite column. Calculate the effective length L_e of the column using the procedure given in the codes.[7.16,7.25]

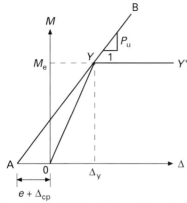

Fig. 7.15 Simplified stability analysis of composite columns

2. Calculate the eccentricity e for the equivalent standard pin-ended column by Eqs 7.27 and 7.28.
3. Estimate a suitable value for the depth of neutral axis d_n. Set either ε_1 or ε_2 (Fig. 7.14) whichever is appropriate, equal to the yield strain of steel ε_y. For these conditions, calculate P_n using Eqs 7.34 and 7.36 to 7.39. When $P_n = P_u/\phi$ where ϕ is the strength reduction factor, the selected value of d_n is acceptable. Otherwise iterate the depth of neutral axis until this happens.
4. For the correct value of the depth of neutral axis, calculate Δ_y using Eqs 7.40 and 7.12. Also, calculate M_n by Eq 7.35.
5. For a known value of P_ϕ, calculate Δ_{cp} by Eqs 7.14, 7.18 and 7.20. Estimate deflections due to initial imperfections in the steel sheath when they are significant and add these to Δ_{cp}.
6. For these values of Δ_y and Δ_{cp}, calculate M_e by Eq. 7.22. The given composite section is considered to be adequate when M_e is less than ϕM_n.

The strengths predicted by the above method show good correlation with the measured values of 50 test columns reported in the literature.[7.8,7.36,7.37] The method is illustrated by an example in Section 7.6.

7.6   Design of Composite Columns – Example

A tubular column filled with concrete has been used as a pile in a marine environment. The diameter of the steel tube $D = 550$ mm with a wall thickness $t = 8$ mm.

Check the adequacy of the pile section for the following data: $P_u = 2225$ kN, $e = 85$ mm, $f'_c = 50$ MPa, $f_y = 350$ MPa, $\varepsilon_y = 0.00175$, $E_s = 200 \times 10^3$ MPa, $E_c = 35\,000$ MPa, $L_e = 13$ m, $P_\phi = 1465$ kN, $I_{gt} = 6820 \times 10^6$ mm⁴, $A_{gt} = 301\,650$ mm², $\phi_{cc} = 2.0$ and $\phi = 0.6$.

Depth of Neutral Axis

After several trials, select the depth of neutral axis $d_n = 305$ mm. Because this value is greater than $D/2$, obviously failure is governed by ε_1. Therefore, set $\varepsilon_1 = \varepsilon_y = 0.00175$. For $d_n = 305$ mm, from geometry of the cross-section, $A_{sc} = 7512$ mm^2, $d_{sc} = 185$ mm, $A_c = 127\,857$ mm^2, $d_c = 127$ mm, $A_{st} = 6103$ mm^2 and $d_{st} = 161$ mm.

From Eqs 7.36 and 7.37, $\sigma_{sc} = 200 \times 10^3 \times 0.00175 \times 185/305 = 212$ MPa and $\sigma_{st} = 200 \times 10^3 \times 0.00175 \times 161/305 = 185$ MPa. Therefore, $C_s = 7512 \times 212 = 1593$ kN and $T = 6103 \times 185 = 1129$ kN. Also, from Eq. 7.39, $\varepsilon_c = 0.00175 \times 127/305$, or $\varepsilon_c/\varepsilon_0 = 0.364$. From Eq. 7.38, $\sigma_c = 0.85 \times 50\,[(2 \times 0.364) - (0.364)^2] = 25.3$ MPa and hence $C_c = 25.3 \times 127\,857 = 3235$ kN. From Eq. 7.34, $P_n = 1593 + 3235 - 1129 = 3699$ kN. $P_u/\phi = 2225/0.6 = 3708$ kN which is close enough to P_n. Therefore, accept the depth of neutral axis as 305 mm.

Calculate Δ_y and M_n

From Eqs 7.40 and 7.12, $\Delta_y = (0.00175/305)(13\,000/\pi)^2 = 98$ mm. Also, the lever arms of forces measured from the plastic centroid are $z_{sc} = 155$ mm, $z_c = 97$ mm and $z_{st} = 191$ mm. From Eq. 7.35, $M_n = (1593 \times 0.155) + (3235 \times 0.097) + (1129 \times 0.191) = 776$ kN m.

Calculate Δ_{cp}

Assume the concrete is uncracked due to sustained load P_ϕ. Then, $P_c = (\pi/13\,000)^2 \times 35 \times 10^3 \times 6820 \times 10^6/2.6 = 5362$ kN, $P_{c0} = 5362 \times 2.6 = 13\,943$ kN. From Eqs 7.14 and 7.18, $\Delta_{tot} = 85/[5362/1465) - 1] = 32$ mm, $\Delta_e = 85/[13\,943/1465) - 1] = 10$ mm.

From Eq. 7.20, $\Delta_{cp} = 32 - 10 = 22$ mm. Assume that deflections due to imperfections in the steel sheath are negligible.

The extreme fibre tensile stress in the concrete $\sigma_1 = (P_\phi/A_{gt}) - P_\phi(e + \Delta_{tot})$ $y_t/I_{gt} = (1465 \times 10^3/301\,650) - 1465 \times 10^3(85 + 32)267/6820 \times 10^6 = -1.85$ MPa (tension). The tensile strength of concrete is $0.6\sqrt{f_c'} = 0.6\sqrt{50} = 4.24$ MPa. Therefore the concrete is uncracked when P_ϕ acts.

Check Adequacy of Section

From Eq. 7.22, $M_e = 2225(85 + 22 + 98)/10^3 = 456$ kN m which is less than $\phi M_n = 0.6 \times 776 = 466$ kN m. Therefore, the pile section is adequate.

It is expected that the top 5 m of the pile may be lost due to corrosion. To compensate this loss, additional reinforcing bars must be provided in this part of the pile so that ϕM_n of the reinforced concrete section is not less than that of the composite section. If we take $f_y = 400$ MPa for reinforcing bars and cover = 65 mm, twelve 32 mm diameter bars are required to meet this requirement.

With the advent of high-strength rolled hollow sections (with yield strengths up to 700 MPa) and high-strength concretes (with compressive strengths greater than 50 MPa), the composite column offers significant economic benefit. Instead of considering concrete and steel as two materials in competition with each other, it is prudent to treat them as complementary. The concrete-filled tubular steel columns have significant potential towards achieving this union between these two major materials used in construction.

References

7.1   ACI Committee 363 1992 *State-of-the-Art Report on High-Strength Concrete* (ACI363R-92), American Concrete Institute

7.2   FIP-CEB 1990 *High-Strength Concrete: State of the Art Report* June, FIP, London

7.3   Choy R S 1988 *High Strength-Concrete* Technical Report No TR/F112, Cement and Concrete Association of Australia, Sydney

7.4   Lloyd N A, Rangan B V 1993 *High-Strength Concrete: A Review* Research Report No 1/93, School of Civil Engineering, Curtin University of Technology, Perth, Australia

7.5   Randall V, Foot K 1989 High-strength concrete for Pacific First Centre. *ACI Concrete International: Design & Construction* **11**(4): 14–16

7.6   Smith G J, Rad F N 1989 Economic advantages of high-strength concretes in columns. *ACI Concrete International: Design & Construction* **11**(4): 37–43

7.7   ACI Committee 363 1991 *High-Strength Concrete* ACI Compilation 17, American Concrete Institute, Detroit

7.8   O'Brien A D, Rangan B V 1993 Tests on slender tubular steel columns filled with high-strength concrete. *Australian Civil Engineering Transactions* **CE 35**(4): 287–293

7.9   Carrasquillo R L, Nilson A H, Slate F O 1981 Properties of high-strength concrete subject to short-term loads. *ACI Journal, Proceedings* **78**(3): 171–178

7.10   Carrasquillo P M, Carrasquillo R L 1988 Evaluation of the use of current concrete practice in the production of high strength concrete. *ACI Materials Journal* **85**(1): 49–54

7.11   Collins M P, Mitchell D, MacGregor J G 1993 Structural design considerations for high-strength concrete. *ACI Concrete International: Design & Construction* **15**(5): 27–34

7.12   Zia P, Leming M L, Ahmad S H 1991 *High-Performance Concretes: A State-of-the-Art Report* Report No SHRP-C/FR-91-103, Strategic Highway Research Program, National Research Council

7.13   Hwee Y S, Rangan B V 1990 Studies on commercial high-strength concretes. *ACI Materials Journal* **87**(5): 440–445

7.14   Moreno J 1991 The state-of-the-art of high-strength concrete in Chicago: 225 West Wacker Drive. *High-Strength Concrete* ACI Compilation 17, American Concrete Institute, Detroit, pp 67–71

7.15   ACI-ASCE Committee 441 1994 High-strength concrete columns: state of the art, Sub-Committee Report, American Concrete Institute, Detroit (private communication)

7.16 Yong Y K, Nour M G, Nawy E G 1988 Behaviour of laterally confined high-strength concrete under axial loads *ASCE Journal of Structural Engineering* **114**(2): 332–351

7.17 Cusson D, Paultre P 1994 High-strength concrete columns confined by rectangular ties. *ASCE Journal of Structural Engineering* **120**(3): 783–804

7.18 Li Bing, Park R, Tanaka H 1994 *Strength and Ductility of Reinforced Concrete Members and Frames Constructed using High-Strength Concrete* Research Report No 94–5, Department of Civil Engineering, University of Canterbury, Christchurch, New Zealand

7.19 Ibrahim H H H, MacGregor J G 1994 *Flexural Behaviour of High-Strength Concrete Columns* Structural Engineering Report No 196, Department of Civil Engineering, University of Alberta

7.20 Rangan B V, Saunders P, Seng E J 1992 *Design of High-Strength Concrete Columns* ACI Special Publication SP-128, American Concrete Institute, Detroit, pp 851–862

7.21 Lloyd N A, Rangan B V 1994 *High Performance Concrete Columns* ACI Special Publication SP-149, American Concrete Institute, Detroit, pp 379–390

7.22 Lloyd N A, Rangan B V 1995 *Behaviour of High-Strength Concrete Columns under Eccentric Compression* Research Report, No 1/95, School of Civil Engineering, Curtin University of Technology, Perth, Australia

7.23 Warner R F, Rangan B V, Hall A S, 1989 *Reinforced Concrete*, 3rd edn, Longman, Cheshire, Australia

7.24 Kaar P H, Hanson N W, Capell H T 1978 Stress–strain characteristics of high-strength concrete. *Douglas McHenry International Symposium on Concrete and Concrete Structures* Special Publication SP-55, American Concrete Institute, Detroit, pp 161–185

7.25 ACI Committee 318 1989 *Building Code Requirements for Reinforced Concrete* (ACI 318-89), American Concrete Institute, Detroit

7.26 Standards Australia 1994 *Australian Standard for Concrete Structures* (AS3600-1994), Standards Australia, North Sydney

7.27 Park R 1993 Application of the New Zealand Concrete Design Code NZS3101 to high-strength concrete, International Workshop on HSC, Kyoto, Japan (private communication)

7.28 Norwegian Standards 1989 *Concrete Structures, Design Rules* NS3473

7.29 Amirthandandan K, Rangan B V 1991 Strength of reinforced concrete columns in biaxial bending. *Civil Engineering Transactions, The Institution of Engineers, Australia* **CE33**(2): 105–110

7.30 Thurairajah S, Rangan B V 1991 Analysis of slender reinforced concrete columns. *The Indian Concrete Journal* **65**(11): 565–570

7.31 Rangan B V 1989 Lateral deflection of slender reinforced concrete columns under sustained load. *ACI Structural Journal* **86**(6): 660–663

7.32 Rangan B V 1990 Strength of reinforced concrete slender columns. *ACI Structural Journal* **87**(1): 32–38

7.33 Webb J, Peyton J J 1990 Composite concrete filled steel tube columns. *Structural Engineering Conference* The Institution of Engineers, Australia, Adelaide, pp 181–185

7.34 Smit E L 1991 The design of concrete filled steel tubular columns. *15th Biennial Conference* Concrete Institute of Australia, Sydney, pp 287–311

7.35 Bruechle P C, Evans E P 1991 Central Park – off-site fabrication/on-site integration. *Proceedings of the International Conference on Steel and Aluminium Structures* Volume on Steel Structures, Singapore, Elsevier Science Publishers, Essex, pp 989–994

7.36 Rangan B V 1991 Design of slender hollow steel columns filled with concrete. *Proceedings of the International Conference on Steel and Aluminium Structures* Volume on Composite Steel Structures, Singapore, Elsevier Science Publishers, Essex, pp 104–112

7.37 Rangan B V, Joyce M 1992 Strength of eccentrically loaded slender steel tubular columns filled with high-strength concrete. *ACI Structural Journal* **89**(6): 676–681

8   The Use of Masonry in Concrete
Buildings

A. W. Hendry

Masonry walls may be used in concrete structures for cladding or for the sub-division of internal space and in some cases are designed to act compositely with the concrete structure. In this chapter discussion of the properties of masonry and masonry materials is followed by consideration of the functions of masonry clad-ding. The requirements for non-loadbearing internal walls are briefly described. The remaining sections of the chapter are devoted to the structural design of infill panels and composite wall beams. A list of key references is provided.

8.1   Introduction

Masonry walls in concrete buildings may be used to provide external cladding or for subdivision of internal space. Such walls are non-loadbearing but will have certain structural requirements, particularly resistance to out-of-plane forces. For certain purposes, however, masonry walls may be designed to act compositely with adjacent concrete elements and will in such cases have to resist in-plane forces.

The essential advantage of masonry lies in the fact that it can meet several functional requirements simultaneously, including weather exclusion, sound and thermal insulation and fire protection, as well as possessing adequate strength. In addition, masonry offers a wide choice of colour and texture and great flexibility in architectural treatment along with durability and relatively low cost.

8.2   Masonry Materials

8.2.1   Masonry Units

Masonry units for cladding and internal walls are available in a wide range of materials and types, as indicated in Table 8.1. Figure 8.1 shows a selection of the commoner shapes and sizes of units, but innumerable configurations are produced for particular applications.

Bricks are typically up to 300 mm in length, 75 mm in height and 100 mm in thickness, although some special types are considerably thicker. Those classed as solid may have indentations on one or both bed faces or may have holes normal to the bed faces to the extent of 20–25 per cent of the area. Above that limit they are classed as perforated. They are made in clay, calcium silicate, and dense and lightweight concrete.

Blocks are available in the same materials but in a much larger variety of shapes and sizes. A common nominal size is 400 mm by 200 mm, length by

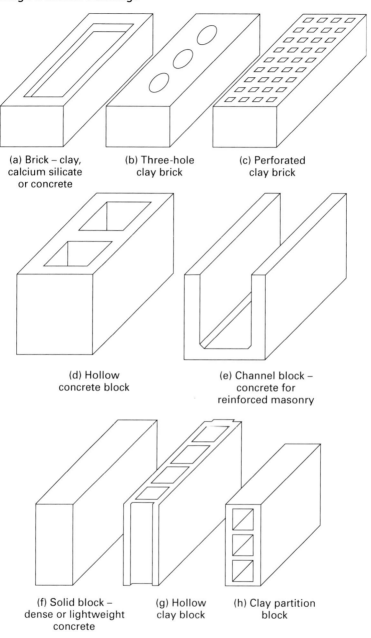

(a) Brick – clay,
calcium silicate
or concrete

(b) Three-hole
clay brick

(c) Perforated
clay brick

(d) Hollow
concrete block

(e) Channel block –
concrete for
reinforced masonry

(f) Solid block –
dense or lightweight
concrete

(g) Hollow
clay block

(h) Clay partition
block

Fig. 8.1 Typical brick and block types

height, and thickness 100 or 200 mm. However, in lightweight materials very much larger sizes are possible. For example, 'jumbo' blocks in aerated autoclaved concrete can be obtained up to 1 m in length and over 600 mm in height, requiring only four courses in a storey height. Blocks can be solid, perforated or

Table 8.1 Masonry units

Material	Unit	Type
Clay	Bricks	Solid
		Perforated
	Blocks	Perforated
		Hollow
Calcium silicate	Bricks	Solid
	Blocks	Perforated
		Hollow
Dense concrete	Bricks	Solid
	Blocks	Solid
		Perforated
		Hollow
		Cellular
Lightweight concrete	Bricks	Solid
	Blocks	Solid
		Perforated
		Cellular

hollow. Provided that the core design of hollow blocks permits them to be in vertical alignment when laid, reinforcement can be placed in the cores, surrounded by concrete. Cellular blocks are those in which the core does not go through the full depth of the unit, giving a full bed face, but much reduced weight as compared to a solid unit.

Compressive strength is the main indicative property for masonry units but may be of less direct importance in relation to non-loadbearing walls than other properties. Thus, the suction rate of clay bricks influences the adhesion between clay bricks and mortar and in turn the resistance to flexure and shear.

Dimensional stability of masonry is important and particular consideration has to be given to movement relative to a concrete structure when used as a cladding to the latter. Clay bricks show some degree of expansion when they take up moisture on removal from the kiln, depending on the type of clay from which they are made and the firing temperature. Most of this irreversible expansion takes place within a short time of manufacture. Reversible movement is also experienced with variation in the moisture content of the unit. Calcium silicate and concrete units also show dimensional changes with moisture content, but with these materials the initial movement is shrinkage. As some movement is associated with mortar, it is more useful to quote these movements for masonry rather than for units and this is done in Section 8.3.4.

8.2.2   Natural Stone

Although less frequently than brickwork and blockwork, natural stone masonry is used as a cladding on concrete buildings, particularly in conservation areas where concordance with existing buildings is important. As distinct from the use of very thin slabs as a facing material, stone masonry may be built as the outer leaf of a cavity wall, the inner leaf of which is in brickwork, blockwork or concrete. Alternatively, a solid wall using stone built against a brickwork backing may be used.

Sandstones or limestones are the most likely to be used for cladding walls. They vary enormously in their characteristics and are chosen mainly on the basis of colour, durability and availability. The compressive strength of sandstones may be in the range 30–90 MPa and limestones 7–70 MPa, but for cladding walls strength in compression is unlikely to be a critical factor. The linear coefficients of expansion and reversible moisture movements for these stones are shown in Table 8.3 (Section 8.3.4), but no information is known for irreversible movement, although it is believed to be very small.

8.2.3   Masonry Mortar and Infill Concrete

For cladding and partition walls in masonry, high strength is generally secondary to avoidance of cracking whether arising from restrained shrinkage, moisture or thermal movements. The use of weak mortar mixes results in some degree of plasticity in the masonry which relieves stresses set up by these effects and thus minimizes cracking.

Mortar mixes suitable for brick and block masonry cladding and internal walls would be as follows:

Cement : lime : sand	Cement : sand with plasticizer	Laboratory strength at 28 days (MPa)
1 : 1 : 6	1 : 3–4	3.6
1 : 2 : 8–9	1 : 5.5–6.5	1.5

For stone masonry a 1 : 3 : 12 cement : lime : sand or crushed stone mix would be suitable. It is important that the texture and colour of the mortar should relate to that of the stone and that dense, impervious mortars should not be used with permeable stone as damage to the latter may result. The content of alkali sulphates in the mortar should be as low as possible to avoid staining and efflorescence or even spalling of the stone by subsurface crystallization.

Masonry walls can be reinforced by placing steel in the cores of blocks, or in the cavity between masonry leaves, surrounded by small aggregate concrete. A suitable mix for this material would be 1 : 0–0.25 : 3 : 2 cement : lime : sand : 10 mm maximum size aggregate. To ensure complete filling and compaction the slump of the mix should be between 75 and 175 mm depending on the height of the pour. Plasticizers or superplasticizers may be necessary for filling small cores or narrow cavities. Particular care is necessary to protect reinforcement against corrosion and in external walls the use of galvanized or austenitic stainless steel will generally be necessary.

8.2.4   Masonry Wall Ties and Fixings

Ties between the leaves of cavity walls are embedded in the horizontal joints of the masonry at a spacing depending on the type of the ties, the leaf thickness and the cavity width. For leaf thicknesses from 65 to 90 mm and cavity widths 50 to 75 mm, ties should be placed at 450 mm centres, horizontally and vertically. If the leaf thickness is greater than 90 mm and the cavity width 50–150 mm the horizontal spacing may be increased to 900 mm, but for a cavity width greater than 75 mm, vertical twist ties should be used. Butterfly ties are not in any case recommended in buildings over two storeys. Additional ties are required at openings in a wall. Stainless-steel ties are to be preferred for external walls as loss of ties by corrosion is a common and serious defect.

Conventional ties are suitable for cavity walls in which the outer leaf is un-supported for a height of up to about 9 m. If a greater unbroken height is required the differential movements should be estimated and special ties used to accommodate them.

Great care must be taken to ensure that ties are correctly spaced and embedded in the joints according to the specification. Deficiencies may only be revealed after a serious and expensive failure.

Fixings are required to provide lateral support to wall panels from the concrete structure. Various types are commercially available, a commonly used form being the dovetail tie, shown in Fig. 8.2. This is suitable for securing a wall to a concrete column or slab, the slot component being cast into the concrete, whilst the actual tie is set in the bed joint of the masonry. Relative vertical movement is accommodated and there is ample tolerance for the exact level of the bed joint. Other fixings are designed to give lateral support at the top of a wall panel. An example of a sliding anchor for this purpose is shown in Fig. 8.2.

The number of fixings required should be calculated with regard to the design wind pressures at the location of the building, the panel dimensions and boundary conditions and the recommended design strengths of the fixings used.

8.3   Properties of Masonry

8.3.1   Compressive Strength

Masonry walls in concrete buildings will in general be non-loadbearing so that compressive strength will not be a primary consideration in selection of materials. However, for completeness, a summary of the factors affecting this property will be given in this section.

The compressive strength of masonry is influenced by the following.

- *Masonry unit*: strength of material, shape of unit.
- *Mortar*: Strength, deformation characteristics, thickness of joint.
- *Construction*: through the wall units, vertical joint parallel to face, cavity wall, stone masonry (ashlar, rubble etc.).

A formula which takes into account most of these variables has been proposed for brickwork and blockwork[8.1] in connection with the draft Eurocode 6:[8.2]

$$f_k = K(\delta f_b)^{0.65} f_m^{0.25}$$

where f_k is the characteristic compressive strength, K is a constant depending on the type of construction, δ is a shape factor for the unit, f_b is the compressive strength of the unit and f_m is the compressive strength of mortar.

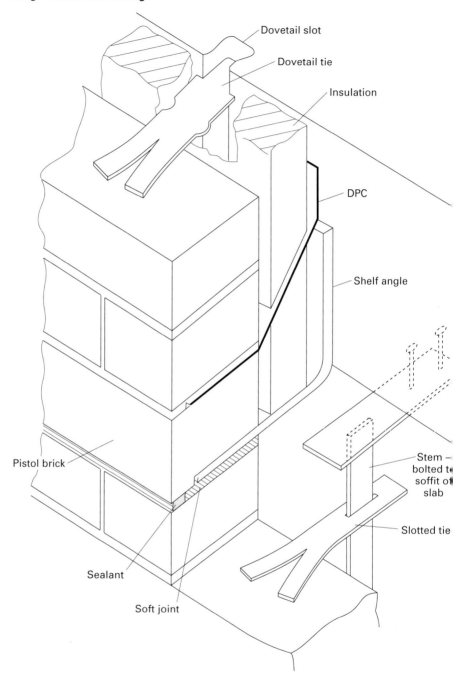

Fig. 8.2 Detail of dovetail and slotted tie fixings between concrete slab and masonry wall

In using this formula in a design code, it is obviously necessary to define the unit and mortar strengths in terms of standardized tests. The shape factor δ applied to the unit strength is intended to give a 'normalized' strength, equivalent to that of a cube of selected size so that a single formula can be used for units of different proportions. The formula does not allow for joint thickness which in modern brick and blockwork is assumed to be around 10 mm. It can, however, be more or less than this in stone masonry, and some masonry recently developed uses either very thin bed joints or even no mortar at all. These variations may be taken into account by selection of a suitable value of the constant K. Possible values of K are as follows.

- For brick or blockwork with solid units equal in thickness to the wall, $K = 0.65$.
- For other units used in single leaves, $K = 0.6$.
- For brick or blockwork where there is a vertical joint parallel to the face, $K = 0.55$.

Values for stone masonry are less well established but 0.55 for ashlar and 0.45 for squared rubble would appear to be reasonable.[8.3]

For regularly shaped blocks the shape factor may be represented by the following formula:[8.4]

$$\delta = (h/\sqrt{A})^{0.37}$$

where h is the height of the unit and A is the loaded area of the unit. This gives the strength of an equivalent 100 mm cube. In the absence of a standard test for stone strength this could reasonably be based on a specimen of this size.

8.3.2   Flexural Strength

Non-loadbearing masonry cladding panels supported by a concrete frame will be required to have sufficient strength to resist wind loading. This is unlikely to be a problem for solid walls of 300 mm thickness or more, but in the case of thinner walls, resistance will be reliant on flexural tensile strength. For clay bricks this strength depends on the water absorption of the unit and the mortar mix, but for concrete blockwork it is related to the compressive strength. Typical values derived from test results[8.5] are given in Table 8.2.

8.3.3   Shear Strength

The shear strength of unreinforced masonry depends on the adhesion between the mortar and the units plus a quasi-frictional increment proportional to the compressive stress normal to the bed joints. If as in a non-loadbearing panel the compression is limited to the self-weight of the masonry the frictional component would be very small and a shear strength of 0.15 MPa would be a conservative assumption for all types of units.

Table 8.2 Flexural strength of masonry (MPa)

		Weak direction		Strong direction	
	Mortar mix:	$1:1:6$	$1:2:9$	$1:1:6$	$1:2:9$
Clay bricks					
Water absorption $<7\%$		0.5	0.4	1.5	1.2
7–12%		0.4	0.35	1.1	1.0
$>12\%$		0.3	0.25	0.9	0.8
Concrete and calcium silicate bricks		0.3	0.2	0.9	0.6
Concrete blocks					
Compressive strength 2.8–7.0 MPa		(a) 0.25	0.2	0.4–0.6	0.4–0.5
		(b) 0.15	0.1	0.25–0.35	0.2–0.3

(a) Wall thickness up to 100 mm.
(b) Wall thickness 250 mm.

The racking shear strength of reinforced masonry walls has been found to depend on the ratio of height to length, the amount and distribution of reinforcement and masonry strength.[8.6,8.7] With reinforcement ratios up to about 0.5 per cent the shear strength of walls in which the reinforcement is placed in pockets, cores or cavities may be taken as 0.7 MPa.

8.3.4   Thermal, Moisture, Elastic and Creep Movements

As mentioned in connection with the properties of units, dimensional changes in masonry resulting from temperature and moisture variations are of importance when that material is used in conjunction with a concrete structure or when two kinds of masonry materials work in parallel as in a cavity wall. Table 8.3 sets out figures for moisture and thermal movements for various types of masonry and for stone, concrete and steel.

Elastic and creep movements may also have to be estimated. An approximate value for the short-term elastic modulus of masonry is $600f_k$ for calcium silicate and AAC brickwork and blockwork and $1000f_k$ for other materials, where f_k is the characteristic compressive strength. In calculating long-term deformations due to creep the short-term modulus should be divided by the factor shown in Table 8.4, according to the material.

Table 8.3 Moisture and thermal movement of masonry, concrete and steel

	Moisture movement (per cent)		
Material	Reversible	Irreversible	Coefficient of thermal expansion/°C (10^{-6})
Clay brickwork	0.02	+ 0.02–0.07	5–8
Calcium silicate bwk	0.01–0.05	− 0.01–0.04	8–14
Concrete bwk or blwk	0.02–0.04	− 0.02–0.06	6–12
AAC blockwork	0.02–0.03	− 0.05–0.09	8
Dense aggregate concrete	0.02–0.10	− 0.03–0.08	10–14
Sandstone	0.07	–	7–12
Limestone	0.01	–	3–4
Steel	–	–	12

Table 8.4 Reduction factors applied to
elastic modulus of masonry for long-term
deformations

	Factor
Clay	1.7
Calcium silicate	2.5
Dense aggregate concrete	2.5
Lightweight aggregate concrete	2.0
Autoclaved aerated concrete	2.5

8.3.5   Partial Safety Factors

Structural design in most countries is now based on limit state principles. Although there are various approaches to this the International Standards Organization (ISO) has formulated a system of partial safety factors,[8.8] some applied to loading and others to material strengths. The latter are essentially intended to allow for the difference between laboratory strengths and site strengths. Variability of material strength is allowed for by adopting the 95 per cent confidence limit on the laboratory strength as the relevant value. This in turn is divided by the partial safety factor, usually designated γ_m, to give the design strength of the masonry. Selection of the appropriate value of γ_m depends on the level of site supervision and of manufacturing control on the units, but will lie in the range 2.0–3.5.

8.4   Masonry Cladding

8.4.1   Cladding Walls: General

Masonry external walls used as cladding on concrete buildings may be of a variety of materials including clay, calcium silicate or concrete brickwork or blockwork, or natural stone, and may be of solid, cavity or veneer construction. Solid walls may be of bonded brickwork or of units of the same thickness as the wall. Cavity walls consist of two leaves or wythes of the same or different materials and thicknesses, connected by corrosion-resistant wall ties. In masonry veneer, widely used in North America, a thin facing of brick is attached to a back-up system of reinforced concrete or steel studs by specially designed, flexible metal ties. Figure 8.3 shows typical examples of masonry cladding walls and Fig. 8.4 is a detail of a masonry veneer system on steel stud backing.

8.4.2   Thermal Insulation

Most masonry walls, unless of lightweight material, require thermal insulation to be incorporated in the wall. This may be a suitable sheet material on the inside of a solid wall, or the holes in perforated or hollow units may be filled with foamed polystyrene or similar. Cavity-wall insulation in new construction can take the form of glass- or rock-fibre slabs or expanded polystyrene sheets placed in the cavity. Existing walls can be insulated by filling the cavity with urea formaldehyde foam or one of a range of loose-fill materials.

(a) Solid bonded brickwork   (b) Cavity wall

(c) Stone masonry outer leaf,   (d) Brickwork veneer
    concrete inner wall   with metal studs

Fig. 8.3 Examples of masonry cladding walls

Care has to be taken in the design of external walls to avoid cold bridges; for example if the floor slabs are brought through the thickness of the wall there is a break in the insulation which increases the heat loss from the building and may result in condensation. In colder climates moist air leaking through a cavity wall may condense on the inside of the outer leaf in sufficient quantity to cause efflorescence on the surface of brickwork. Air leakage also results in significant heat loss and is prevented either by non-permeable insulation or by a separate vapour barrier. In either case, sealing should be provided wherever there is a break in continuity of this layer.

8.4.3   Avoidance of Rain Penetration

Perhaps the most complex problem in the design of masonry cladding walls is to ensure that there is no penetration of rainwater to the interior of the building. Two aspects have to be considered – first, the resistance of the wall itself and, second,

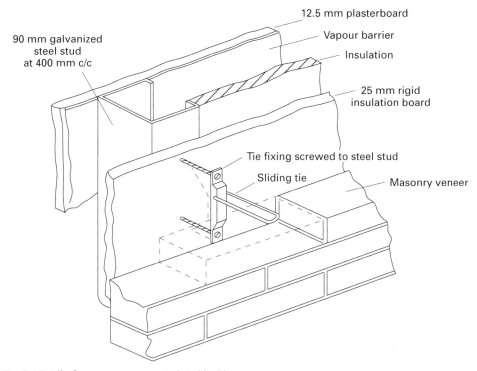

90 mm galvanized
steel stud
at 400 mm c/c

12.5 mm plasterboard

Vapour barrier

Insulation

25 mm rigid
insulation board

Tie fixing screwed to steel stud

Sliding tie

Masonry veneer

Fig. 8.4 Detail of masonry veneer on steel stud backing

leakage at openings and around the perimeter of the wall. Information on the selection of wall types and details for particular situations is given in codes of practice and design guides[8.9,8.10] and only an outline can be given here.

Considering the resistance of single-leaf masonry walls to rain penetration, the minimum thickness necessary to give satisfactory service depends on the exposure, the type of masonry and whether it is rendered. The standard of workmanship and maintenance also influence the performance of a wall. Architectural features, such as string courses, which throw water off the surface of the masonry reduce the likelihood of rain penetration. Conversely, the lack of such features and the presence of large glazed areas draining on to masonry increase the chances of water getting through the wall. In unfavourable circumstances of exposure, design and construction, there is no practical thickness of solid masonry which can be assumed watertight.

The definition of exposure is related to a combination of wind and rainfall and should be assessed in accordance with a recognized method such as the British Standards Institution publication DD 93.[8.11] This gives five grades of exposure varying from 'Very severe' to 'Very sheltered'. Associated with these categories the British Standard BS 5628 Part 3 recommends masonry thicknesses for rendered and unrendered, clay, calcium silicate and concrete masonry. For very severe conditions the use of single-leaf masonry is not recommended at all and, unless thicknesses in excess of 190 mm are used, unrendered masonry would only be suitable for very sheltered locations. If the outer face of the masonry is rendered, a 215 mm thickness would be acceptable for moderate/severe conditions. It is nevertheless possible to use 102.5 mm thick clay facing brickwork in stud cladding systems in conjunction with suitable measures to prevent water which inevitably penetrates the masonry skin from entering the building.

The limitations of single-leaf masonry in relation to rain penetration have led to the widespread adoption of the cavity wall. In such a wall water will penetrate the outer leaf, but with correct design and construction will not cross the cavity. The cavity should be at least 50 mm in width and if cavity insulation is used the least probability of rain penetration will be achieved with slab type insulant fastened to the face of the inner leaf with adhesive. Cavity-fill insulants of the urea formaldehyde type carry an increased risk of allowing water to cross the cavity, but of course must be used in the case of existing walls.

The most common causes of leakage arise from careless workmanship allowing mortar to accumulate in the cavity or from the incorrect installation of damp-proof courses. Damp-proof courses are required at the base of each section of wall and weepholes must be provided at these locations to drain any water which may have penetrated the outer leaf. A typical detail at a column/floor-slab junction is shown in Fig. 8.5. Damp-proof courses are also necessary around door and window openings and at the base and head of parapet walls.

8.4.4   Provision for Differential Movement

Cladding walls may be supported on shelf angles or projections from the main structure at intervals up the height of the building. Alternatively, they may be built independently of the main structure but tied to it for lateral support. In both cases provision for relative movement between the masonry wall and the concrete structure is a primary design requirement. Thus it is common practice to limit the

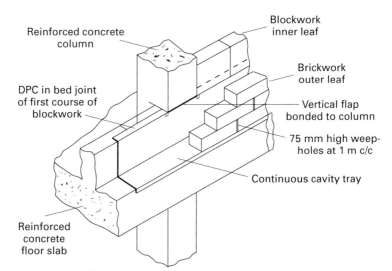

Fig. 8.5 Typical detail of cavity wall damp proof course at column/floor slab junction

uninterrupted height of the outer leaf of a cavity wall to three storeys or 9 m. This limit may be increased to four storeys or 12 m in buildings not exceeding this height or number of storeys.[8.5]

If these limits are to be exceeded the relative movement between the masonry and the concrete structure must be estimated and the construction detailed to accommodate the expected movements. These will result in the masonry from shrinkage, elastic deformation, creep, thermal and moisture effects. A conventional allowance for the movement to be expected in clay brickwork is an expansion of 1 mm per metre. Aerated concrete blockwork, on the other hand, shrinks by up to 1.2 mm per metre. These movements have to be compared with the expected movement in the concrete structure to which the walls are to be attached.

The result of a detailed assessment of the vertical movement of the outer, stone masonry leaf of a cavity wall having an uninterrupted height of 25 m is shown in Table 8.5. The assumptions made are as follows.

- Unit weight of masonry: 2400 kg m^{-3}.
- Shrinkage of stone masonry: 0.04 mm m^{-1}.
- Elastic modulus of masonry: 10.8 MPa.
- Creep of masonry: 1.5 times elastic deformation.
- Coefficient of thermal expansion: $10 \times 10^{-6}\,°C^{-1}$.
- Temperature at construction: 10°C.
- Maximum temperature $+40°C$; minimum temperature: $-30°C^{-1}$.
- Reversible moisture movement: $+0.4$ mm m^{-1}.

The elastic and creep deformations are calculated storey by storey on the basis of the dead-load stresses and summed to give the cumulative compressions up the height of the building in which the upper five storeys were 3.15 m in height and the lower two 4.45 and 4.7 m respectively. It will be noted that the elastic deformation only applies to ties after they are laid since that deformation in the wall below the level of a tie will have taken place prior to that. It is necessary to assume a temperature at construction and also whether the units are wet or dry at the time of laying. In the UK the Building Research Establishment recommend a temperature of 10°C for the former. The reversible moisture movement can be plus or minus according to the condition at the time of laying and can therefore be assumed to be in the least favourable direction.

Table 8.5 Relative wall tie movements at each storey level

Storey	Movement (mm)					
	Shrinkage	Elastic	Creep	Thermal	Moisture	Total
7	− 1	0	− 12	− 8	− 10	− 31
6	− 1	− 0.3	− 11	− 6	− 8	− 26
5	− 0.7	− 0.5	− 11	− 5	− 7	− 24
4	− 0.5	− 0.7	− 10	− 4	− 5	− 20
3	− 0.3	− 1	− 9	− 3	− 4	− 17
2	0	− 2	− 7	− 2	− 2	− 13
1	0	− 3	− 7	− 1	− 1	− 12

Having estimated the movement of the masonry leaf, this must be compared with the corresponding estimate for the concrete structure. In the usual situation the inner leaf of the cavity wall will be built, storey by storey, between the floor slabs of the main structure and will thus move with it. Suitable ties and fixings will then be selected having regard to the relative movements. As the differential movements will be smaller at the lower levels, it will probably be unnecessary to use special ties for the first few storeys.

If a single leaf or both leaves of a cavity wall are used as an infill in a concrete frame, horizontal movement joints will be required at the top of the masonry panels, sufficient to allow for shrinkage of the frame and, if clay brickwork is used, expansion of the masonry. If such a detail is omitted, the result may be bowing of the masonry panels between floors or displacement of the slips covering the floor slabs (Fig. 8.6(a)); detail shown in Fig. 8.6(b) will avoid this occurrence.

Provision must also be made for horizontal movements in the form of vertical movement joints. These will be somewhat less than those in the vertical direction since elastic and creep deformations will be reduced and there will be frictional constraint at the base of the wall. It is normal practice to space vertical-movement joints not more than 15 m apart in clay brickwork, 9 m calcium silicate and 6 m in concrete masonry. These joints are filled with a compressible material and their

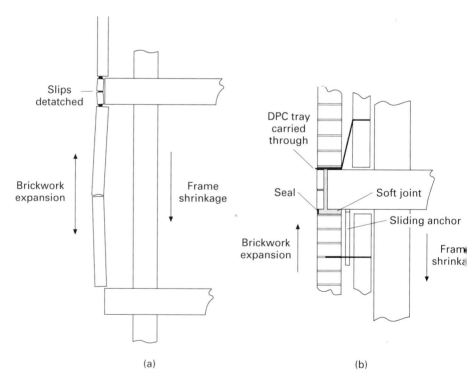

(a)   (b)

Fig. 8.6 (a) Bowing of infill wall and attachment of brick slips as a result of frame shrinkage; (b) detail of horizontal movement joint to avoid damage as in (a)

width in millimetres should be about 30 per cent more than their spacing in metres. Their location will depend on features of the building such as columns in the main structure and re-entrant corners.

It should be noted that the type of mortar used has an important influence on the ability of masonry to accommodate movement. Thus a stone masonry wall in weak lime mortar can be of great length without showing signs of cracking. Brickwork built in strong modern mortar, on the other hand, will have a much lower tolerance of movement and the provision of movement joints will be essential.

8.4.5   Lateral Resistance

External masonry walls in concrete frame buildings must be able to withstand lateral forces arising from wind and from any additional loads for example from material stacked against them. There is no entirely satisfactory method for calculating the strength of laterally loaded panels although a considerable amount of research on the problem has been carried out in the UK and in Australia. A method based on the use of yield line formulae for panels without openings is given in the British code BS 5628 Part 1[8.5] and extended in an appendix to deal with walls with openings. Based on this and on particular wind loads, tabulated values of limiting areas of cavity walls and 190 mm thick solid walls are given in Part 3 of the same code.[8.9]

The design moment per unit height of a panel is taken to be:

$$\alpha W_k \gamma_f L$$

when the plane of failure is perpendicular to the bed joints or

$$\mu \alpha W_k \gamma_f L$$

when the plane of failure is parallel to the bed joints, where α is a bending moment coefficient, γ_f is a partial safety factor for loads, μ is the orthogonal strength ratio, L is the length of panel between supports and W_k is the characteristic wind load per unit area.

The design moment of resistance of a masonry wall is given by

$$f_{kx} Z / \gamma_m$$

where f_{kx} is the characteristic flexural strength appropriate to the plane of bending, γ_m is the partial safety factor for the material and Z is the section modulus.

Equating the design moment to the moment of resistance gives the minimum thickness of a solid wall for the given loading as

$$t = \sqrt{\frac{6 \gamma_m \alpha \gamma_f W L}{f_{kx} b}}$$

where b is the width of the wall. The coefficient α is obtained for the given ratio of height to span of the wall, orthogonal ratio and the boundary conditions from the appropriate yield line formulae. These are to be found in various reference books[8.12,8.13] or in BS 5628 Part 1. The code also gives values of the characteristic flexural strength of various types of masonry derived from tests on small panels.

Cavity walls can be calculated by this method assuming that the strength of the wall is equal to the sum of that of the two leaves, provided that the wall ties are capable of transmitting the load between the leaves.

In addition to giving a method for the calculation of the strength of laterally loaded panels in relation to specified wind loads, BS 5628 places the following limits on the dimensions of such walls.

1. Panel supported on three sides:
    (a) two or more sides continuous – height × length < 1500 t;
    (b) all other cases – height × length < 1350 t.
2. Panel supported on four sides:
    (a) three or more sides continuous – height × length < 2250 t;
    (b) all other cases – height × length < 2025 t.
3. Panel simply supported top and bottom: height × length < 40 t.

The provisions described above refer essentially to unreinforced masonry walls, but could be applied to reinforced masonry if the appropriate moment of resistance is used. The flexural behaviour of reinforced masonry sections is in principle the same as that of reinforced concrete[8.14] and for rectangular sections is represented by the equation

$$\frac{M_d}{bd^2} = \rho f_y \left(1 - \frac{0.5\rho f_y}{f_m} \frac{\gamma_{mm}}{\gamma_{ms}}\right) < 0.4 \frac{f_m}{\gamma_{mm}}$$

where M_d is the moment of resistance, b is the width of the section, d is the effective depth of section, ρ is the steel ratio, f_y is the characteristic strength of steel, γ_{ms} is the partial safety factor for steel, f_m is the characteristic strength of masonry and γ_{mm} is the partial safety factor for masonry.

The term on the right-hand side of this equation represents a cut-off for compressive failure of the masonry. Figure 8.7 shows a family of design curves based on the above equation and particular values of the partial safety factors and steel strengths. The compressive failure cut-off appears as a diagonal line and a second cut-off for shear failure is shown as a dashed line.[8.14]

Bed-joint reinforcement, whilst limited in the area of steel which can be introduced, is known to increase the lateral strength of a single-leaf wall and will have the effect of altering the orthogonal ratio in the yield line method. In hollow blockwork walls there will be greater scope for the introduction of vertical steel in the cores and in some cases horizontal steel. This, however, is unlikely to be necessary solely for resistance to wind loads unless in very large panels.

8.5   Internal Walls

Internal walls may be for subdivision of space within a building with single occupation or to form party walls between different occupiers. Partition walls

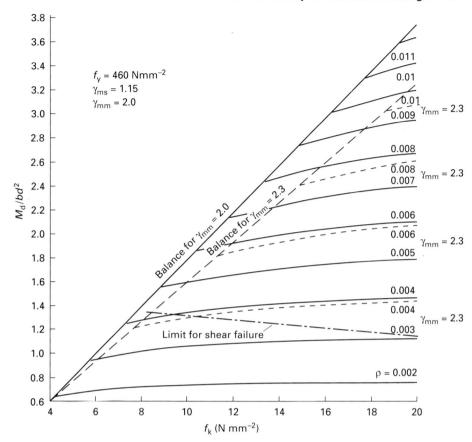

Fig. 8.7 Design graph for reinforced masonry

require to be reasonably robust, provide prescribed fire protection and sufficient sound insulation for privacy. The requirements for party walls are more stringent both in respect of fire protection and sound insulation. For the latter, mass is important, and if a masonry party wall is adequate in this respect fire protection is likely to centre on detailing to prevent smoke and flame from spreading round the perimeter of the wall.

Internal walls in masonry are not usually designed for specific loading conditions, but in order to ensure adequate strength and stability certain limits on length and height relative to thickness are necessary. Such limits are recommended in BS 5628 Part 3[8.9] and are shown in Fig. 8.8.

It is also necessary to ensure that partition walls are not damaged by deflection of concrete members above or below which they are built. If a masonry wall is supported by a concrete beam or slab it may be necessary to allow for deflection of the latter by providing vertical movement joints. Alternatively, a separation joint at the base of the wall would allow independent movement of the concrete element but may require bed-joint reinforcement in the wall where tension might

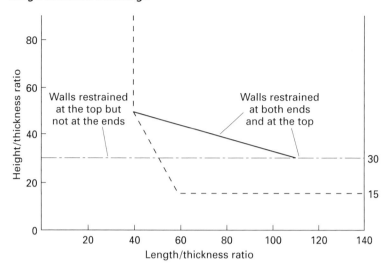

Fig. 8.8 Limiting dimensions of internal walls for stability (BS 5628 Part 3 1985)

develop. If a masonry partition wall is built up to the underside of a concrete member, either a gap should be left at the wall head or a layer of resilient material should be placed there to prevent damage to the masonry as a result of deflection of the slab or beam.

8.6   Masonry Infill Panels in Concrete Buildings

8.6.1   Infill Walls: General

It is common practice to infill reinforced concrete frames with masonry walls. These may form either external or internal walls and are not usually assumed to contribute to the strength of the frame. However, numerous studies have shown that concrete frames are substantially stiffened and strengthened under lateral loading by masonry infill, and this cannot be neglected in considering seismic response.

If infill panels are not intended to carry vertical loading it is necessary to provide horizontal compressible joints at the head of each section of wall to allow for shrinkage and creep of the concrete structure and possible expansion of the masonry if this is clay brickwork. Calcium silicate and concrete masonry will shrink so that the differential movement in this case will be smaller. Relative movement should also be allowed on the vertical edges of panels while still providing support.

8.6.2   Structural Effect of Infill Walls

In the previous section it was assumed that no structural interaction was intended between the concrete frame and the masonry infill panels. It may, on the other hand, be the intention of the designer to make use of the stiffening effect of the masonry in resisting lateral loads, and in this case methods are required for the calculation of such actions. A considerable amount of research has been reported on this problem and is reviewed in Refs 8.15 and 8.16. An obvious difficulty which arises in applying the various theories which have been developed arises from the practical necessity, discussed in the preceding paragraph, to allow for differential movement between the frame and the infill. This implies that there is unlikely to be full interconnection between the frame and the masonry panel. Although this may add to the analytical difficulties,[8.17] it is a feature of infilled frame behaviour that separation between the elements generally takes place before the ultimate load is reached and that racking shear loads are transferred into the masonry at diagonally opposite corners of the panel through lengths of masonry which depend on the stiffness of the frame, as suggested in Fig. 8.9(a). Failure of the infill takes place either by the development of cracks or by local crushing at the loaded corners (Figs 8.9(a) and (b)). Wood has produced equations which apply to various failure modes[8.17] and has suggested a factor to adjust the effective crushing strength of the masonry in the light of experimental results.

An approximate method of calculating the strength of an infilled panel is to estimate the proportion of the applied load carried by the infill using the graph shown in Fig. 8.10.[8.16] This is then compared with the shear strength of the infill panel calculated on the basis of the horizontal cross-section and the shear strength of the masonry, whether plain or reinforced.[8.18] It is also necessary to check the strength of the masonry against local crushing. There are theoretical methods for calculating the contact length between frame members and infill, but these are of somewhat doubtful accuracy because of the uncertain initial fit of the infill. It is therefore advisable to make a conservative estimate of the length of contact and ensure that the masonry is of sufficient compressive strength to transmit the racking load without premature failure. A number of tests on frames with concrete block infill[8.19] suggested that the corner blocks should be assumed to carry this load and designed accordingly.

8.6.3   Infilled Frames in Seismic Conditions

Important qualifications apply to the structural response of infilled frames to seismic forces.[8.18] Thus the stiffening effect of even quite weak masonry infill panels can attract forces to parts of the structure which may not have been designed to resist them. Such effects are exaggerated if the infill panels are unsymmetrically disposed in an otherwise symmetrical frame since then torsional moments will result which may cause additional damage. There is also the danger of shear damaged masonry falling from the structure, especially if it had been attempted to isolate the panels structurally from the frame. It follows that if infill masonry is used in seismic-resistant structures it will be necessary to consider the disposition of the panels and to design the frame and the walls on the assumption of integral behaviour. Reinforced masonry construction will generally be required which will minimize the danger of walls being displaced by out-of-plane seismic forces.

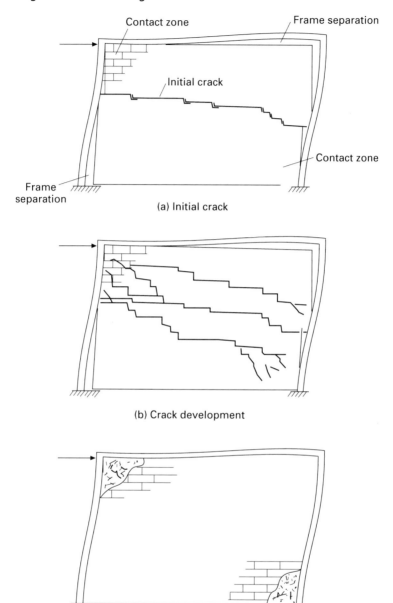

(c) Crushing at loaded corners

Fig. 8.9 Typical failure patterns in frame with brickwork infill

8.7   Composite Concrete/Masonry Beams

8.7.1   Masonry Walls Supported by Concrete Slabs or Beams

Where a masonry wall is supported by a reinforced concrete slab or beam, it has long been recognized that the whole weight of the wall is not supported by the

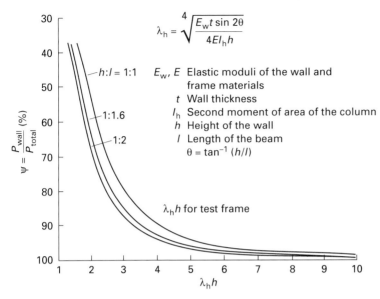

$$\lambda_h = \sqrt[4]{\frac{E_w t \sin 2\theta}{4EI_h h}}$$

E_w, E Elastic moduli of the wall and
frame materials
t Wall thickness
I_h Second moment of area of the column
h Height of the wall
I Length of the beam
$\theta = \tan^{-1}(h/I)$

$\lambda_h h$ for test frame

Fig. 8.10 Proportion of load carried by masonry in infilled frames

concrete member as a uniformly distributed load. It is therefore common practice to consider the loading on the beam to be confined to a triangular section of the wall, as indicated in Fig. 8.11, resulting in a much reduced bending moment. This is a considerably simplified representation of what takes place if the masonry is laid on a mortar bed on the concrete member so that there is composite action between the two.

It is possible to design masonry to act compositely with reinforced concrete beams or slabs to produce a lintel or a wall beam. A considerable amount of relevant research has been carried out on composite construction and various design methods have been developed.

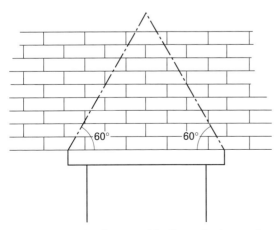

Fig. 8.11 Conventionally assumed loading on lintel supporting masonry wall

8.7.2   Calculation of Moments and Forces in Composite Beams

The structural action of a composite wall beam is as indicated in Fig. 8.12. Vertical and shear forces are concentrated towards the supports resulting in a kind of arching action in the wall. A convenient solution for the analysis of this system is available[8.20] based on a parameter

$$R_f = \sqrt{\frac{E_m t H^3}{EI}}$$

where E_m is the elastic modulus of masonry, t is the thickness of the wall, H is the height of the wall and EI is the flexural rigidity of the beam.

This is a measure of wall-to-beam stiffness. The ratio of the maximum to the average compressive stress in the wall, derived by finite-element calculations is given by

$$f_m/(W_w/Lt) = (1 + \beta R_f)$$

where f_m is the maximum compressive stress in the wall, W_w is the total vertical load on the beam, L is the span, t is the thickness of the masonry and β is a coefficient.

Assuming a triangular distribution of vertical stress between the wall and the beam, the contact length is thus

$$L_v = L/(1 + \beta R_f)$$

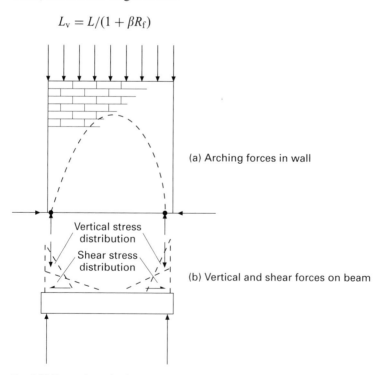

(a) Arching forces in wall

Vertical stress distribution

Shear stress distribution

(b) Vertical and shear forces on beam

Fig. 8.12 Composite action between masonry wall and reinforced concrete beam

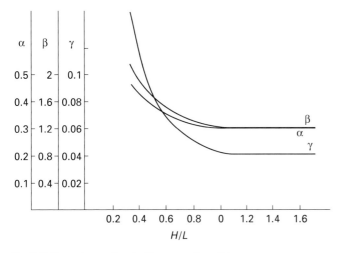

Fig. 8.13 Parameters α, β and γ for composite wall beams

The horizontal force in the beam is transmitted from the wall by shear at the interface and is given by

$$T = W_w(\alpha - \gamma R_a)$$

where α and γ are further coefficients, $R_a = E_w tH/(EA)$ and A is the area of the beam. Values of the coefficients α, β and γ are shown in Fig. 8.13.

Finite-element analysis has shown that the shear stress acts over a length two to three times that of the vertical stress, that is

$$L_s = 2L(1 + \beta R_f)$$

Again assuming a triangular stress distribution, the maximum shear stress is

$$V_m = [W_w(\alpha - \gamma R_a)(1 + \beta R_f)]/Lt$$

The bending moment at any section of the beam results from a combination of moments due to the vertical loading and the shear force at the interface of the wall and beam. The bending moments in the beam depend on the parameters R_f and R_a.

Investigations[8.21] have indicated that centrally placed openings in a wall beam have relatively small effect on bending moments in the beam but a large opening at about quarter span will double the bending moment.

An alternative approach which appears to be suitable for calculating the required steel area over the full range of beam height-to-span ratios[8.22] is based on a modification of the basic equations for the design moment of resistance of a reinforced masonry beam:

$$M_d = A_s f_y d[1 - 0.5 A_s f_y \gamma_{mm}/(bd f_k \gamma_{ms})]/\gamma_{ms}$$

and

$$M_d < 0.4 f_k bd^2/\gamma_{mm}$$

where A_s is the steel area, f_y is the characteristic yield strength of steel, γ_{ms} is the partial safety factor for steel, d is the effective depth, b is the breadth, γ_{mm} is the partial safety factor for masonry and f_k is the characteristic compressive strength of masonry.

Assuming that the beam is simply supported and carries a uniformly distributed load of w kN m^{-1} then $M_d = wl^2/8$. Introducing the two non-dimensional parameters

$$A = A_s f_y/(\gamma_{ms} wL)$$

and

$$B = 4\gamma_{mm}w/(bf_k)$$

the basic equations can be rewritten as follows:

$$A = [(4kH/L) - \sqrt{(4kH/L)^2 - B}]/B$$

and

$$B < 12.8k^2(H/L)^2$$

where $k = d/H$, i.e. the ratio of the effective to the overall depth.

In order that the square-root term in the equation for A is positive it can be shown that B must be less than $16k^2(H/L)^2$ which requirement is automatically satisfied if the criterion for avoiding compressive failures is met.

This solution is shown graphically in Fig. 8.14, assuming the limiting value of $B = 12.8k^2(H/L)^2$ and $k = 0.8$. For a given H/L, the corresponding value of A can be read off and the required steel area found from the equation defining this parameter.

In addition to determining the steel area required in a composite beam, it will be necessary to check that the local compressive stresses in the vicinity of the supports do not exceed the design compressive strength of the material.

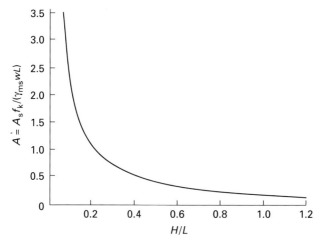

Fig. 8.14 Variation of parameter A with ratio H/L for composite masonry/reinforced concrete beams

References

8.1   Edgell G J 1989 The characteristic compressive strength of masonry. *Masonry International* **3**(1): 6–11

8.2   Commission of the European Communities 1988 *Common Unified Rules for Masonry* Eurocode No 6, EUR 9888

8.3   Hendry A W 1993 Assessment of stone masonry strength in existing buildings. *Proceedings 5th International Conference on Structural Faults and Repair* ECS Publishers, Edinburgh

8.4   Khalaf F M, Hendry A W 1978 Shape factor for masonry units. *Proceedings British Masonry Society* No 5

8.5   British Standards Institution 1978 *Code of Practice for the Use of Masonry* BS 5628 Part 1

8.6   Schneider R R, Dickey W 1980 *Reinforced Masonry Design* Prentice Hall, pp 528–563

8.7   Scrivener J C 1969 Static racking tests on concrete masonry walls. *Designing, Engineering and Constructing with Masonry Products* Gulf Pub Co, pp 24–31

8.8   International Standards Organization 1973 *General Principles for the Verification of the Safety of Structures*, ISO 2394

8.9   British Standards Institution 1985 *Code of Practice for the Use of Masonry* BS 5628 Part 3

8.10 Plewes W G, Coakeley M J 1981 *Exterior Wall Construction in High Rise Buildings*, Canada Mortgage & Housing Corp, Ottawa

8.11 British Standards Institution undated *Methods for Assessing Exposure to Wind-driven Rain* DD93

8.12 Roberts J J *et al.* 1983 *Concrete Masonry Designers' Handbook* Viewpoint Publishers

8.13 —— 1987 *Safe Load Tables for Laterally Loaded Masonry Panels to BS 5628 Part 1* Kenchington Little Publishers

8.14 Hendry A W 1991 Reinforced masonry elements in flexure *Reinforced and Prestressed Masonry* (ed Hendry A W), Longman, pp 60–73

8.15 Hendry A W 1990 *Structural Masonry*, Macmillan, pp 231–243

8.16 Scrivener J C 1991 Reinforced masonry shear walls and buildings under static load. *Reinforced and Prestressed Masonry* (ed Hendry A W), Longman, pp 177–180

8.17 Liauw T C, Kwan K H 1983 Plastic theory of non-integral infilled frames *Proceedings of the Institute of Civil Engineers*, Part 2, **75**: 379–396

8.18 Priestley M J N 1991 Seismic design of reinforced masonry. *Reinforced and Prestressed Masonry* (ed Hendry A W), Longman, pp 186–189

8.19 Hendry A W 1993 Tests on steel frames with reinforced masonry infill. *Proceedings British Masonry Society* No 5

8.20 Davies S R, Ahmed A E 1978 An approximate method for analysing composite wall/beams. *Proc B Ceram Soc* **27**: 308–320

8.21 Davies S R, Ahmed A E 1976 Composite action of wall-beams with openings. *Proceedings Fourth International Brick and Masonry Conference* Brugge, Paper 4.b.6

8.22 Davies S R, Hendry A W 1991 A comparison between the design of masonry beams. *Proceedings Asia-Pacific Conference on Masonry* Singapore, pp 31–36

9   Precast Concrete Skeletal Structures

Kim S. Elliott

Precast concrete offers a versatile alternative to cast in situ concrete for large multistorey structures. Early consultation with the precast engineer can result in cost savings, improved buildability and shorter construction times. Prefabrication offers a large potential for the future: production line economy combined with quality assurance both at the factory and on site ensures the highest specification structures.

This chapter discusses aspects of the design, manufacture and construction of precast skeletal structures. These buildings consist essentially of beam–column structures, with prestressed long span flooring and shear cores and/or walls placed mainly around lift shafts and staircases. Skeletal structures are used for buildings of up to 15 to 20 storeys depending on ground conditions and stability provisions, and clear floor spans of up to 20 m are possible using prestressed concrete components. Diaphragm action can be achieved without a structural topping screed on the floor. Most connections are designed and installed as pinned joints to prevent complex erection methods leading to errors. Aspects of robustness in the event of accidents are discussed.

9.1   Current Attitudes Towards Precast Concrete Structures

The latest generation of multistorey precast concrete frames has evolved over the past 25 years over which time the rise in quality and technical achievement has been unrivalled in the concrete industry. One of the largest concrete structures in the world – a 5000-room hotel, leisure complex and car park for MGM in Las Vegas, USA – was constructed during 1992 using precast concrete. Although precast can offer the client structural and architectural freedom, the industry is still labouring under the misconceptions of modular precast buildings, thus:

> ... the design of a precast concrete structure on a modular grid. The grid should preferably have a basic module of 0.6 m...,[9.1]

or

> ... modulation of the grid is often governed by the width of floor and facade units ... 600, 1200 or 2400 mm....[9.2]

A clear distinction needs to be made between 'modular coordination' and standardization. Modulation offers zero flexibility off of the modular grid. Interior architectural freedom is possible only in the adoption of module quantities and configuration, and one cannot escape the geometrical dominance and lack of individuality of these buildings. Exterior façades may be varied indefinitely, as in the twin façade system which requires a full precast perimeter wall behind a full height façade. Standardization is quite different from modulation. It refers to the manner in which a set of predetermined components are used and connected. Most of the buildings shown in Fig. 9.1 were constructed using the same family of standardized components. By adjusting beam depths, column lengths, wall positions etc. the same components in any of these buildings could have been used to make a completely different structure.

The three basic types of precast structures are as follows.

1. The *wall* frame, Fig. 9.2, consisting of vertical wall and horizontal slab units only, and used extensively for multistorey hotels, retail units, hospitals, housing and offices. The structural walls serve also as partitioning. See Ref. 9.3 for further methods of analysis.

2. The *portal* frame, consisting of columns and roof rafters, and used for single-storey retail warehousing and industrial manufacturing facilities.

3. The *skeletal* structure, Fig. 9.3, consisting of columns, beams and slabs for medium-rise buildings, with a small number of walls for high-rise. Skeletal frames are used chiefly for schools, commercial offices and car parks.

This chapter is concerned only with *skeletal* frames. These are the most architecturally and structurally demanding because in both disciplines, designers feel that they have free rein to exploit the structural system by creating large uninterrupted spans whilst reducing structural depths and the extent of the bracing elements. This results in a large proportion of the connections being highly stressed and difficult to analyse.

The skeletal structure is distinguished from other types because imposed gravity loads are carried to the foundations by beams and columns, and horizontal loads by columns and/or walls. The volume of structural concrete in the beam-column skeleton is of the order of 1–2 per cent of the volume of the building. For a cross-wall frame the figure is closer to 4–5 per cent. Flooring is the most significant factor with prestressed concrete voided slabs adding a further 4 per cent, compared with 7–8 per cent for solid concrete slabs of the same load-carrying capacity.

One of the key issues is programming the deliveries to site so that the fixing team is not under pressure to construct hastily, nor to fix the components out of sequence. This could impair the temporary stability of the structure as the height to the centre of the mass of the concrete above the level at which the frame is stable could be prohibitive to further progress. The rule is that components should not be fixed more than *two* storeys ahead of the last floor to be fully tied into the stabilizing system. This allows time for the *in situ* concrete at the lower levels to mature.

The market for precast structural frames varies throughout the world. This can be as little as 10 to 15 per cent, e.g. in the United Kingdom, where structural steelwork is a major competitor, or as much as 80 per cent in Scandinavia where there is a short season for cast-in-place concrete and where steelwork is imported. The production of precast concrete flooring in Scandinavia approximates to about 0.7 m^2 per capita per year, compared with a European average of about 0.1 m^2 per capita per year.

Fig. 9.1 Possible variations in precast concrete skeletal frames (courtesy Crendon Structures Ltd, Blatcon Ltd, Trent Concrete Ltd. UK)

Fig. 9.1c, d and *e*

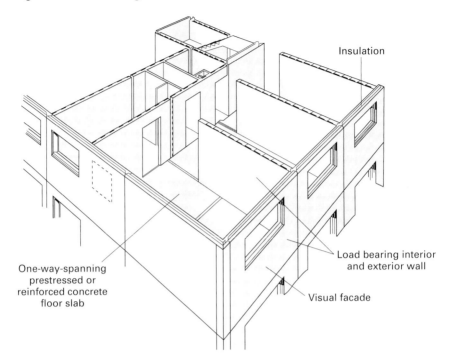

Insulation

One-way-spanning
prestressed or
reinforced concrete
floor slab

Load bearing interior
and exterior wall

Visual facade

Fig. 9.2 Wall frame, incorporating solid (or sandwich) wall panels and voided prestressed floor slabs

9.2   Precast Design Concepts

Huyghe and Bruggeling[9.4] state that '... Prefabrication does not mean to "cut" an already designed concrete structure into manageable pieces...'. The correct philosophy behind the design of precast concrete multistorey structures is to consider the frame as a *total entity*, not an arbitrary set of elements each connected in a way that ensures interaction between no more than the two elements being joined. Thus it is clear that all the aspects of component design and structural stability are dealt with simultaneously in the designer's mind. The main aspects at the preliminary stage include:

1. structural form;
2. frame stability and robustness;
3. component selection;
4. connection design.

These items *cannot* be dealt with in isolation. For example, the nature of the column–beam connection dictates the arrangement and function of the reinforcement in the ends of beams, and the manner in which floor slabs are connected to edge beams influences the torsional behaviour of the beam. It must also be remembered that two additional procedures, namely manufacture and site erection, are also directly influential in making design decisions.

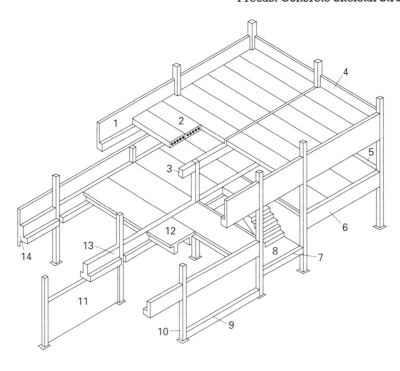

1 Main spandrel beam	8 Staircase and landing
2 Hollow-core unit	9 Ground beam
3 Internal rectangular beam	10 Column
4 Gable spandrel beam	11 Wall
5 Gable beam	12 Double-tee unit
6 Main edge beam	13 Internal beam
7 Landing support beam	14 Main edge spandrel beam

Fig. 9.3 Definitions of components in a skeletal structure

The first task is to establish an economical plan layout for the optimization of the minimum number of the least-cost components versus overall building requirements. The optimum is usually found in a rectangular grid where the beams span in a direction parallel with the greater dimension of the frame. The flooring should span on to the beams or walls making an angle of intersection of not less than 45° to the direction of span. Primary columns are located at the strategic points (corners, changes in floor level, around stairwells and lift shafts) and secondary columns are introduced to satisfy architectural requirements, or to obtain structural economy by using the minimum number of components giving acceptable structural zones.

Precast manufacturers have standardized their components by adopting a range of preferred cross-sections for each type of component. A single method of connection (bolted, doweled or welded) should be used throughout a structure. The rules for determining whether a structure is to be braced or unbraced are fairly clear cut, and they affect both prefabricated steel and concrete frames alike. Table 9.1 gives guidance. Further explanations are given later in the chapter.

Table 9.1 Storey heights of precast concrete frames and types of bracing

Approximate range for number of storeys	Bracing element(s)
Unbraced frame	
$\leqslant 3$	Columns
Braced frame	
$\leqslant 4$	Steel cross-bracing
$\leqslant 5$	Brick or block infill wall
3–6	Precast concrete wind posts (deep columns)
3–10	Precast concrete infill walls
3–12	Precast concrete hollow core shear walls
10–15	Precast concrete shear boxes
15–20	*In situ* concrete shear core
Partially braced frame*	
5–10	Precast concrete infill walls
5–12	Precast concrete hollow core

9.3   Precast Superstructure Simply Explained

9.3.1   Basis for the Design

The superstructures shown in Figs 9.1 and 9.3 are designed with 'pin-jointed' connections between columns, beams and floor slabs. This structure is braced using strategically positioned shear walls or cores. In all cases horizontal wind loads are transmitted through the precast floor, including unscreeded slabs, as though the floor were a deep beam. The chord elements to this deep beam are the precast beams themselves, and therefore the perimeter of the frame must be capable of carrying the tension and compression continuously as shown in Fig. 9.4.

One-way spanning prestressed (or reinforced) precast floor slabs are recessed into beams. A structural topping screed is seldom used in office buildings of seven to 10 storeys or less, depending on the plan dimensions and distances between shear walls. Beams are connected to columns and walls using connectors designed as pinned joints (Fig. 9.5). Fully-rigid frame connections are possible in certain situations where a moment of resistance can be generated at the beam-to-column connection, but semi-rigid methods are not yet sufficiently advanced to be considered at present. An eccentric loading is applied to the column and the bending moment is distributed in the column according to simplified 2D frame analysis. Columns are considered continuous at floor joints, even though mechanical connections, called splices, are often made at this level. The design of connections will be dealt with in Section 9.6.

Structural stability is *the* most crucial issue in precast concrete design because it involves both the design of the precast concrete components and of the connections between them. Precast systems are scrutinized by checking authorities more for structural stability, integrity, resistance against abnormal loading and robustness, than for the design of individual precast components (slabs, beams

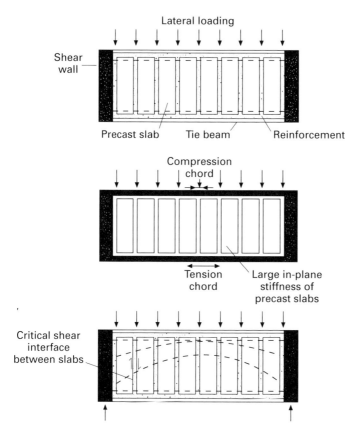

Fig. 9.4 Horizontal floor diaphragm action using discrete precast floor units

etc.) which usually have adequate factors of safety. In general 'stability' means adequate resistance against side sway, and 'integrity and robustness' means correct joint design, attention to details and the prevention of progressive collapse. The problem is to ensure adequate ultimate strength and stiffness, and to ensure that the failure is ductile.

Two design stages are considered as follows.

1. *Temporary stability during frame erection.* This has certain implications on design, e.g. the axial load capacity of temporary column splices (i.e. before the *in situ* grout is introduced to render the splice permanent) has to be greater than the self-weight of the upper column.
2. *Permanent stability.* This may be subdivided into four further stages:
   (a) horizontal diaphragm action in the precast floor slab;
   (b) transfer of horizontal loading from the floor slab and into the vertical bracing elements and the foundations;
   (c) component design;
   (d) joint design.

Fig. 9.5 Beam–column connections are classed as pinned joints

The contribution to the lateral strength and stiffness of the structure from the *in situ* reinforced concrete infill strips between the precast elements is paramount. These strips provide the necessary tie forces which eliminate relative displacements between the various parts of the frame and ensure interaction between the components, for example between beam and slabs, and the slabs themselves. Engineers need to be cautious in allowing service openings or using novel types of connections in these highly sensitive areas.

9.4   Materials

Prefabrication of reinforced and prestressed concrete elements has much greater potential for economy, structural performance and durability than has cast *in situ* concrete. Typical values for the standard deviation of compressive cube strengths are in the order of 4 N mm^{-2} for mean design cube strengths of $f_{cu} = 60 \text{ N mm}^{-2}$. The 28-day strengths and short-term secant elastic moduli (N mm^{-2}) for the range of standard mixes used are given in Table 9.2.

One of the main mix design parameters is the strength of concrete at an age of 24 hours for beam, column, wall and staircase units, and between 12 and 24 hours for prestressed hollow cored slabs. This is the optimum age when the units are removed from the mould, and the strength at this point in time is termed the lifting strength. For prestressed units it coincides with the de-tensioning strength. Recycled concrete aggregates are used by some producers to replace up to 8–9 per cent of the virgin material. Pozzolans, such as premium grade (best grade to national standards) pulverized fuel ash or blast furnace slag, are used as partial cement replacements, in some cases by as much as 75 kg m^{-3}.

Table 9.2 Typical design strengths and moduli used in precast elements (N mm^{-2})

Component	Type*	Grade	Cube (cylinder) strength	Design strength	Tensile strength	Elastic modulus
Beams, staircases, floors, shear walls	RC	C40	40.0 (30.0)	18.0	–	28 000
Columns,† loadbearing walls	RC	C50	50.0 (40.0)	22.5	–	30 000
Beams, floors, staircases	PSC	C60	60.0 (50.0)	27.0	3.50	32 000

* RC: reinforced concrete; PSC: prestressed concrete.
† Very long columns may be prestressed to aid transportation and pitching operations.

9.5   Precast Concrete Flooring Options

9.5.1   Types and Comparisons

Precast concrete flooring offers an economic and versatile solution to ground and suspended floors in any type of building construction. The main types of flooring used in multistorey construction are shown in Fig. 9.6 as (a) hollow cored slabs, (b) double tee slabs and (c) beam and composite plank. Table 9.3 summarizes the performance criteria for each.

More than 90 per cent of all precast concrete used in flooring is prestressed. The slabs are manufactured and designed in accordance with national codes and other selected literature produced by Sheppard and Phillips,[9.5] Elliott,[9.6] BCA,[9.7] I.Struct.E,[9.8] PCI,[9.9] FIP[9.10] and individual research.[9.11]

9.5.2   Hollow-Cored Units

The units are manufactured using the slipforming (non-circular voids) or long line extrusion (circular voids) process. Steel beds, of very high accuracy, are used in lengths of up to 150 m × 1.20 m wide (or multiples or fractions thereof). Slabs nearly 3.6 m wide were specially manufactured close to the site of the aforementioned MGM building. Recent advances in robotics have opened the way for automatic void forming during manufacture, and 1994 saw the introduction of 500 mm deep units, adding larger spans to an impressive specification.

The slabs are designed as a series of I sections with narrow webs. The concrete is grade C50 minimum. The section is structurally very efficient and the design is well documented in the FIP Manual.[9.10] The maximum span-to-depth ratio is about 55 (lightly loaded floors or roofs), and the holes guarantee the maximum possible radius of gyration in addition to reducing the self weight of the slab by 40 to 53 per cent.

The slabs are reinforced longitudinally using 9 mm to 15 mm diameter seven-wire helical strand, or 5 mm to 7 mm crimped or plain wire. Owing to the absence of stirrups, the minimum flange thickness must be at least 30 mm. The

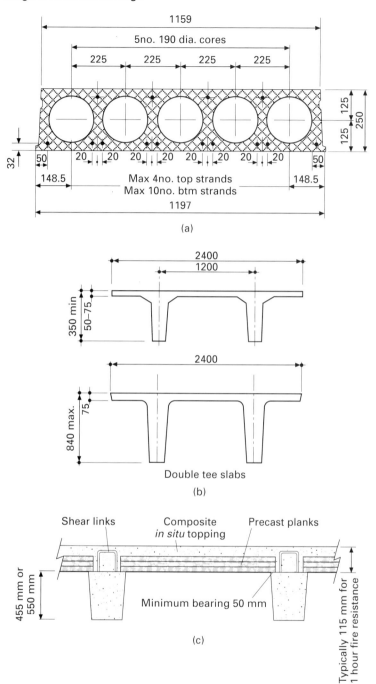

Fig. 9.6 Types of precast concrete floors

Table 9.3 Properties and performance characteristic of precast flooring

Type	Usual widths (mm)	Typical depths (mm)	Approx. self-weight (kN m^{-2})	Approx. maximum span* (m)
Hollow core	1200, 600, 400, 333	110	1.90	7.0
		150	2.25	8.5
		200	3.00	10.5
		250	3.35	11.0
		300	3.60	14.5
		400	4.80	18.5
		500	5.50	22.5
Double tee	2400, 3000	400	3.60	14.5
		500	4.10	17.0
		600	4.50	19.5
		700	4.95	21.0
		840	5.50	24.0
Composite beam and plank	900–2400 beam centres	455/115† 550/115†	4.2–6.6‡ 4.5–7.4‡	17.8 20.9

* Span for superimposed loading of 1.5 kN m^{-2} plus finishes of 1.5 kN m^{-2}.
† Depth of precast beam/depth of plank and topping.
‡ Depends on centres of precast beam elements.

minimum web thickness depends on the overall depth of the unit h, and is given by $1.6\,h^{0.5}$. The rules for the size of the longitudinal gap between adjacent units are shown in Fig. 9.7. They are based on the premise that tie steel might be placed in the joint. Edge profiles have evolved to ensure an adequate transfer of horizontal and vertical shear between adjacent units using structural grade *in situ* concrete or grout (C25 minimum).

9.5.3 Double Tee Slabs

Typical cross-sections are shown in Fig. 9.6(b). The vast majority of units are prestressed, although reinforced concrete units are equally feasible. These units are cast in steel moulds – the depth of the web may be varied by placing block-outs in the crevices in the mould. The strength of the units is in the two deep webs

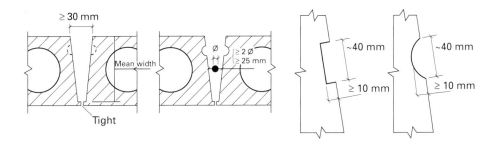

Fig. 9.7 Detailing of the longitudinal joint between hollow-cored floor slabs

where prestressing tendons are laid in pairs. The main advantages in using this type of unit (compared with the hollow-cored units) are:

- load-carrying capacity, i.e. greater loads at the same span;
- the ends of the units can be notched to form a halving joint to reduce the overall structural depth;
- the units are manufactured up to 2400 mm wide (actually 2390 mm) – thus reducing the number of units to be fixed on site, although greater capacity cranage may be required.

Due to the shallow flange depth – between 50 mm and 75 mm thick – an *in situ* reinforced concrete structural screed is required to ensure vertical shear transfer between adjacent units and horizontal diaphragm action in the floor plate.

9.5.4   Precast Beam and Plank Flooring

The beam and plank floor is commonly known as the PBC system, and is shown in Fig. 9.6(c). It is a three-part construction, as follows.

1. The prestressed beams are the major structural elements and are seated on to bearing pads, typically 150 mm square by 10 mm thick. The spacing between the beams varies between 900 mm and 2400 mm.
2. The precast planks, which are generally 75 mm to 100 mm deep solid section, span transverse to the beams.
3. A 40 mm to 50 mm thick grade C30 to C40 *in situ* concrete topping is added to the top of the planks, making the total construction span/depth, inclusive of all three items, of the order of 25 : 1 for offices, and 30 : 1 for domestic buildings.

Voids may easily be formed between the prestressed beams, up to about 2.0 m wide. The system is more versatile geometrically than either the hollow core or double tee slabs and may be adapted for a wide range of plan configurations, as shown in Fig. 9.8.

9.6   Design of Frame Components

9.6.1   Beams

Edge beams are subjected to predominantly asymmetrical loads where the floor is carried by a projecting boot, thus creating an L-shaped section. The upstand may or may not form a part of the structural section as shown in Fig. 9.9 as either:

- Type I, where the upstand is part of the structural section; or
- Type II, where the upstand provides a permanent formwork to the floor slab, and is ignored in the basic design stage for Type IIa, but included in the final beam design stage for Type IIb.

Minimum depth is often determined by the size of the connector in the end of the beam. In this case the minimum depth would be equal to the depth of the floor

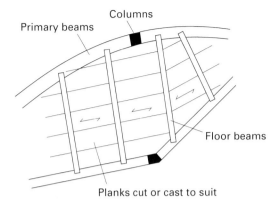

Fig. 9.8 Versatility of the beam and plank flooring method

slab (h_s) plus the minimum boot depth of 150 mm. The boot is designed on the principle of a continuous nib where the horizontal component of the floor load V is $H = V \tan \theta$, where θ is the angle formed between the compressive diagonal strut beneath the load point and the horizontal tie steel in the top of the boot, and θ should not be less than about 30°. The horizontal bars placed in the top of the boot must satisfy $A_{sh} = H\gamma_{m(steel)}/f_y$. The bars are formed into links, but do not contribute to vertical shear unless the boot is sufficiently deep that the vertical leg of the boot link extends beyond the neutral axis of the section. Punching shear forces and lateral bursting forces are determined where point loads, such as double tee supports, are present. Additional boot links (say two 10 mm dia. bars at 75 mm centres either side of the support) are occasionally required for the punching shear requirements in shallow boot depths of say 200 to 250 mm.

Edge beams subjected to asymmetrical loading are not always reinforced against torsion on the assumption that the floor plate provides a horizontal prop force and the lateral stiffness of the beam is sufficiently large to prevent excessive horizontal deflections under the action of the propping force (see Fig. 9.10). However, some engineers dispute this point of view, particularly if the floor loading and/or floor span is large, as is often the case where double tee slabs are

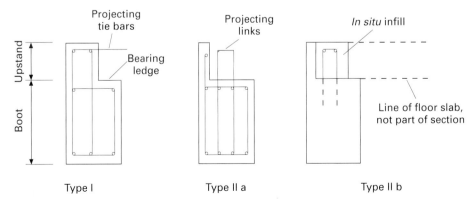

Fig. 9.9 Types of edge L-section beams

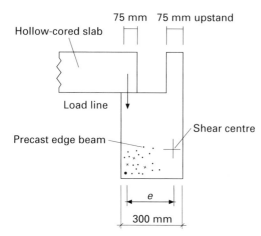

Fig. 9.10 Equilibrium torsion in edge beams subjected to asymmetrical loads

specified. It is more difficult to justify the elimination of torsion using the sole propping action of the structural screed because of the problems in establishing a practical shear transfer mechanism in this type of construction.

An example of this concerns the design of edge beams subjected to asymmetrical loading as shown in Fig. 9.11. It would appear that the beam should be designed against the torque $T=wLe/2$. The value of T can be of the order of 30 to 40 kN m, resulting in a torsional stress (for a typical 300 × 600 mm deep beam) of around 1.5 to 2.0 N mm^{-2} for which torsional links would normally be required. However, edge beams are not reinforced against torsion on the assumption that the precast concrete floor slab, which is not monolithic with the beam but is only tied into the beam, will provide a horizontal propping force to prevent beam rotations occurring.

In the temporary construction stage the beam is not tied to the slab, and therefore only the friction between the floor slab and bearing ledge of the beam will prevent beam rotations. However, the torque induced from the self-weight of the floor slab is only about one-third of the total torque in service, which would not be great enough to cause torsional cracking. The evidence for this, given by Elliott *et al.*[9.12] is as follows.

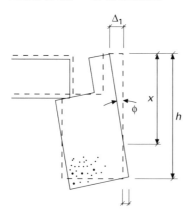

Fig. 9.11 Torsion in L-shaped edge beams

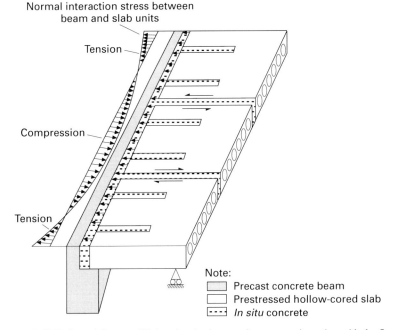

Normal interaction stress between
beam and slab units

Tension

Compression

Tension

Note:
▢ Precast concrete beam
▢ Prestressed hollow-cored slab
⌐---⌐ *In situ* concrete

Fig. 9.12 Horizontal force equilibrium in edge beams where composite action with the floor slab is considered as contributing to the torsional resistance of the beam

1. The *in situ* concrete infill produces an extended bearing, thus reducing the eccentricity of the load with respect to the shear centre of the precast beam.
2. Composite action alters the position of the shear centre to a point near to the centre of the beam, thus reducing the eccentricity even more.
3. The beam cannot possibly rotate because of the rigid plate stiffness of the precast floor slab, causing the horizontal interface stresses shown in Fig. 9.12.

9.6.2   Beam End-Shear Design

Special attention is given to shear reinforcement near to the end connections. Stirrups and bent-up reinforcing bars are provided to ensure the transfer of shear in the critical region. The design methods shown in Fig. 9.13 are similar in principle to the halving joint or corbels. The dashed lines represent principle stress trajectories as determined using a finite-element solution. The concrete provides the compressive force, providing it is restrained against later bursting by the binding steel, and tie bars provide the tensile force resistance. Martin and Korkosz[9.13] cite 147 references (up to 1982), and many aspects of connection design are given in the PCI manual on connections.[9.14]

Steel plates (or angle sections) cast into the beam are sized to ensure that the average bearing stress in the concrete does not exceed $0.6f_{cu}$. The total eccentricity from the face of the column allows 10 mm to 15 mm site fixing tolerances

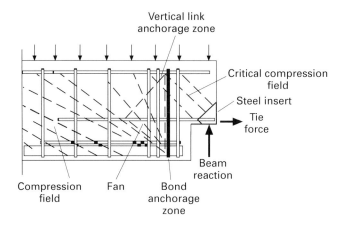

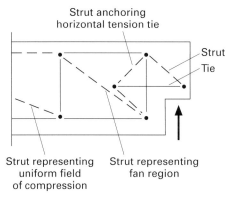

Fig. 9.13 Truss idealization for beam end shear (after Cook[9.18])

between beam and column. The width of the bearing plate b_p should be not less than $b/3$, to avoid punching through the concrete above the plate, and not greater than $(b - 60\text{ mm})$, to give adequate cover with tolerances to the sides of the beam, where b is the lesser of the breadth of the column or beam.

Two methods of design are used as shown in Figs 9.14(a) and 9.14(b). The shear resistance V calculated from each of the two methods are additive, although Ref. 9.13 suggests that when these two modes of action are adequately catered for the next diagonal plane of failure occurs further away from the half joint, and therefore the full vertical shear force resistance must be mobilized well into the main body of the beam. Reinforcement normal to the potential cracking plane provides the normal force required to maintain the ultimate shear resistance. The force $T = \mu V$ is generated because of volumetric changes in the concrete due to shrinkage, temperature and creep etc. F_t is the axial force in the beam. The horizontal reinforcement is given by:

$$A_h = \frac{F_t + \mu V}{f_y / \gamma_{m(\text{steel})}}$$

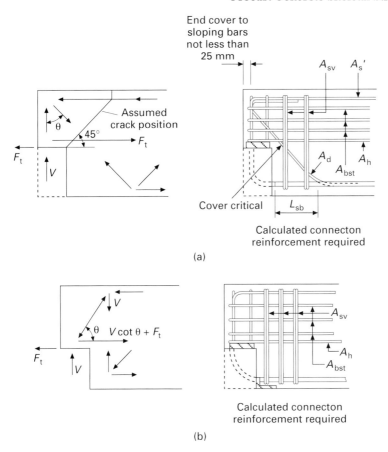

Fig. 9.14 Structural mechanisms and reinforcement details for beam end half joint

where m is the shear friction coefficient, equal to 0.7 for plain concrete-to-concrete surfaces, and 0.4 between steel and concrete, or steel to steel. At least two bars should be used for A_h, and they should preferably be welded to the bearing plate because it is unlikely that full anchorage would be achieved by other means. The area of diagonal bars A_d is given by

$$A_d = \frac{V \sec\theta}{f_y/\gamma_{m(steel)}}$$

where θ is usually 40° to 45°. The minimum bending radius for bent bars must not be contravened because the concrete is already in a 'high risk' zone and localized effects must not be allowed to exacerbate the situation. For this reason 20 mm dia. bars are considered to be the largest practical bar size for A_d.

The reaction V gives rise to a lateral horizontal bursting force H_b which may be obtained from end block theory – the most appropriate technique – to give

$H_b = \eta V$, where z is the bursting force coefficient, typically 0.20 to 0.23 for $b_p/b = 0.2$ to 0.4, respectively. Then

$$A_{bst} = \frac{H_b}{f_y/\gamma_{m(steel)}}$$

To complete the shear cage the first full depth vertical links A_{sv} designed using the truss action, shown in Fig. 9.14(b), should be placed at *one* cover distance from the inside face of the pocket. It is likely that pairs of links are required. The second set of links should be provided within the *nodal* distance L_{sb}.

An alternative to the *shear-cage* concept outlined above is the *shear-box* approach in which a solid plate, RHS or other, structural steel section projects from the end of the beam (see Fig. 9.15). Shear capacity of the section is based on the shear capacity of the shear box itself, and is gradually transferred into the reinforced concrete beam. Tie-back forces are distributed into the concrete beam either by an appropriate concentration of vertical stirrups or by welding a wide plate (or similar) to the bottom of the shear box. Economy is possible only because the steel sections are restrained against distortion by the concrete, and the permissible confined concrete stress may be taken as $0.8f_{cu}$.

Figure 9.15 shows the design method when using a narrow plate insert based on the so-called Cazaly hanger, developed by Lawrence Cazaly of Ontario, Canada. The narrow plate is compatible with the welded plate connector (see Fig. 9.16(b)). See Refs 9.4, 9.6, 9.9 (Section 2.4) and 9.13 for further design information.

9.6.3   Beam-to-Column Connections using Steel Inserts

A wide range of connections, some of which are shown in Fig. 9.16, have evolved to satisfy these requirements. These connections have formed the basis of extensive experimental, numerical and analytical studies (see for example Refs 9.15 to 9.18). The options are as follows.

- Direct frictional bearing between beam and column inserts with no positive mechanical action between the essential precast components (Fig. 9.16(a)). Shear capacity is based *only* on the net shear section of the projecting inserts.
- A welded connection. Top fixings may be excluded because of the torsional stability provided by the weld and a pair of stability pins (Fig. 9.16(b)). The vertical shear capacity excludes the contribution of the weld.
- A separate intermediate cleat using a rolled tee, angle or fabricated plates (Fig. 9.16(c)). Top fixings may be excluded because torsional stability is provided by the bolts (at least two).
- A notched steel plate placed into a box cast into the end of the beam (Fig. 9.16(d)).

In all instances the gap between the precast components is concreted using a grade C30 to C40 sand–cement grout, sometimes containing a proprietary expanding agent to counter shrinkage.

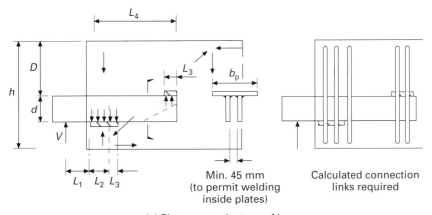

(a) Plate near to bottom of beam

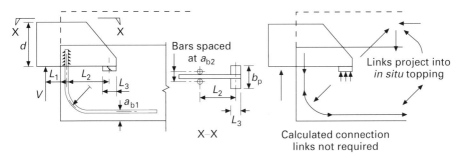

(b) Plate near to top of beam

Fig. 9.15 Narrow plate connector for end shear design[9.8]

Column insert design may be subdivided into either the design of solid inserts (Figs 9.16(a), (b)) or cast-in sections (Figs 9.16(c), (d)). The design of solid inserts varies between *wide sections*, i.e. when the breadth of the bearing surface b_b is in the range 75 mm $< b_b < 0.4b$, and *thin plates* which include thin-walled rolled sections with wall thickness less than $0.1b$.

In wide sections an ultimate bearing stresses of $0.8f_{cu}$ is possible by reinforcing against bursting, spalling and splitting etc. The concrete is confined directly above and below the column insert using closely spaced binders (links). See Refs 9.6 and 9.8 for further design details. The steel sections obtain some support against distortion from the concrete in which they are embedded, but the projecting parts must be checked to ensure that they will not distort excessively nor buckle under the applied joint loads and rotations.

The maximum compressive force occurs below the insert and this gives rise to a horizontal bursting force from which the area of confinement steel is determined. The binders are typically two or three 10 or 12 mm dia. bars at

50 mm centres. Additional capacity can be obtained by welding on reinforcement to the sides of the insert.

9.6.4   Columns

Precast column design is no different to the design of ordinary reinforced concrete columns and walls once all the aspects of manufacture, different types of structural connections and temporary stability have been resolved. Columns have been prestressed axially to enable very long units (up to 25 m) to be pitched without (flexural) cracking. The columns (and walls) are manufactured horizontally. The standards of control are therefore greater than in vertically cast *in situ* work and congested arrangements of reinforcement, particularly at column *splices* and foundations, can be specified with confidence in the knowledge that full compaction of concrete and correct spacing of bars will always be achieved. It

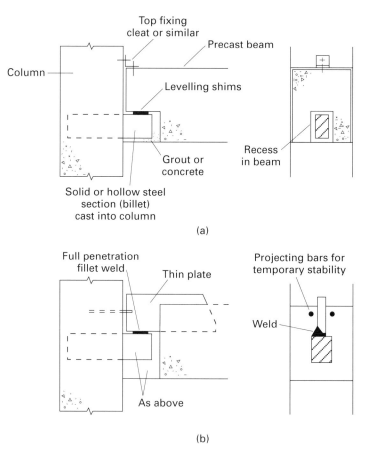

Fig. 9.16 Variations of beam–column connections

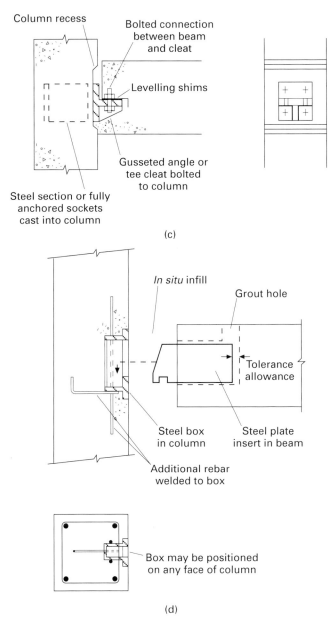

Column recess

Bolted connection
between beam
and cleat

Levelling shims

Gusseted angle or
tee cleat bolted
to column

Steel section or fully
anchored sockets
cast into column

(c)

In situ infill

Grout hole

Tolerance
allowance

Steel box
in column

Steel plate
insert in beam

Additional rebar
welded to box

Box may be positioned
on any face of column

(d)

Figure 9.16c and *d*

is also possible to precast a concrete column having up to 10 per cent reinforcement, although this quantity of reinforcement is rarely used in preference to a larger cross-section. The design strength of the concrete is usually $f_{cu} = 50$ N mm^{-2}, but, because of the early strength required for lifting in the factory, actual characteristic strengths are in the range 60 to 70 N mm^{-2}.

Column bending moments (up to 100 kN m) are the result of eccentric loading in the connection, which varies with the type of connection. Larger bending moments arise from horizontal wind loading, lack of verticality and second-order deflection-induced moments. The resulting moments preclude the use of column splices and pinned bases unless the frame is braced.

9.6.5   Column Splices

A column *splice* is the general term for a joint where a vertical structural connection is made between a column and another precast component. The base member is usually a column, but it may also be a wall, structural cladding panel, beam, or in extreme circumstances a flooring unit. It does *not* include the connection to bases or other foundations.

Designers prefer to stagger the level of column splices at different floor levels to avoid forming a 'plane' of weakness. The level of the first splices is usually shared between the third and fourth floor, except in five-storey frames where the splices (if used at all) are made at the second and third floors. Splices are located either at a floor level (within the structural floor zone) where they may be concealed in the floor finishes, or at a convenient working height, e.g. 1.0 m, nearer to the point of contraflexure in the frame.

Column-to-column splices are made either by coupling, welding or bolting mechanical connectors anchored into the separate precast components or by the continuity of reinforcement through a grouted joint. Figure 9.17 shows the various options. The welded plate and grouted sleeve methods are undoubtedly the most common. See Refs 9.6, 9.8, 9.9, 9.14 and 9.19 for further details.

9.6.6   Column–Base Connections

The design of column connections to pad footings and other *in situ* (or precast) concrete foundations (e.g. retaining wall or ground beam) is well documented.[9.6,9.8,9.9,9.14] Two main methods are used, in pockets or on plates.

Placing columns in pockets is the most economical solution from a precasting point of view, but its use is restricted to situations where fairly large *in situ* concrete pad footings can easily be constructed (Fig. 9.18 and Ref. 9.20). Pocket depths of 1.5 times the column dimension h are used for resisting overturning moments. For pinned bases the pocket depth may be shallower, although $1.5h$ is usually specified. The pocket is usually tapered $5°$ to the vertical to ease the placement of grout in the annulus.

The precast column requires only additional links to resist bursting pressures generated by end bearing forces using $\zeta = 0.11$, and a chemical retardant applied to the mould to help expose the aggregate in the region of the pocket. Vertical loads are transmitted to the foundation by skin friction and end bearing. However, if overturning moments are present half of the skin friction is conservatively ignored due to possible cracking in the precast/*in situ* boundary. The design strength of the grout is usually $f_{cu} = 40$ N mm^{-2}.

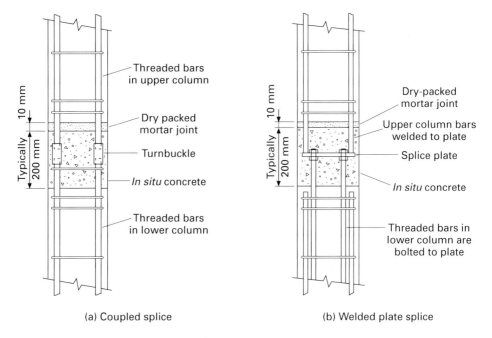

(a) Coupled splice

(b) Welded plate splice

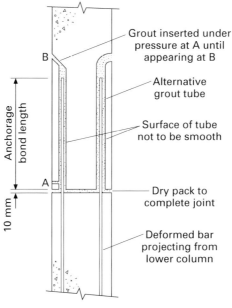

(c) Grouted sleeve splice

Fig. 9.17 Column-to-column splices

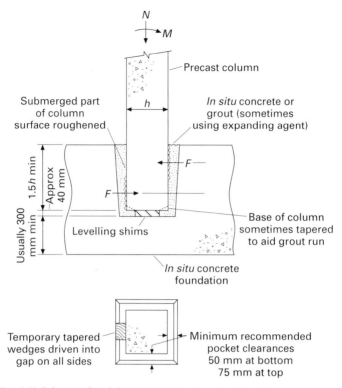

Fig. 9.18 Column-to-foundation base design using grouted pockets

Columns on base plates offer immediate site fixity and are favoured by most frame erectors. Base plates larger than columns are used where a moment connection is required. The maximum projection of the plate is therefore usually restricted to 100 mm, irrespective of the size of the column. Figure 9.19 shows the components involved in this connection. Reinforcement is fitted through holes in the base plate and is fillet welded at both sides. Additional links are provided close to the plate (as is also the practice at splices). Tensile forces due to overturning moments must be transmitted by bond in the precast column, bending and shear in the base plate, tension in the cast-in holding down bolt and bearing and shear in the foundation. Holding down bolts of grade 4 : 6 or 8 : 8 are used as appropriate.

9.6.7   Design of Shear Walls

Infill Shear Walls

Infill shear walls rely on composite action with the pin-jointed column and beam structure for their strength and stiffness. Where an infilled wall is built solidly but not monolithically into a flexible structure its resistance to horizontal loading increases considerably due to composite action with the structure. This is shown in the load response sequence in Fig. 9.20. The problem is similar to stiff beams

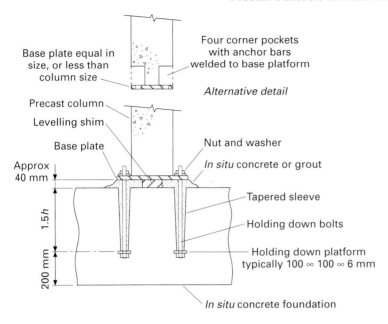

Fig. 9.19 Column-to-foundation design using base plates

on elastic foundations, in that resistance to horizontal loading is enhanced with deformation. An important factor is the quality of the shear key between the wall and frame components, which for manufacturing purposes is usually unreinforced.

Most of the pioneering work on infilled frames – albeit using masonry infill – was carried out by Stafford-Smith and Carter,[9.21] Mainstone[9.22] and Wood.[9.23] Kwan and Liaum[9.24] extended the analysis to the ultimate limit state. The design assumption is that ultimate horizontal forces are resisted by a compressive *diagonal strut* across the concrete infill wall. The effective width of the strut depends primarily on the relative stiffnesses of wall panel and structure, and on *h*

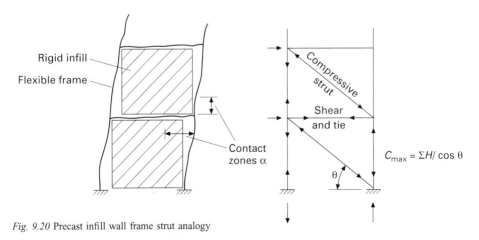

Fig. 9.20 Precast infill wall frame strut analogy

and $\theta = \tan^{-1} h'/l'$, as defined in Fig. 9.21, but it becomes smaller as the interface cracking load is exceeded. Slender wall panels are designed as slender braced *plain* concrete walls taking into account manufacturing and site inaccuracies and deflection-induced bending moments. The shear resistance at the horizontal interface between beam and wall panel is also based on shear in plain concrete walls (normal aggregate, grade C25 minimum) under compression.

It is necessary to calculate the contact lengths α between frame and wall (Fig. 9.20), and find an effective width for the equivalent strut (Fig. 9.21). The ultimate strength of the wall will be given as R_v. The contact length α is given by

$$\frac{\alpha}{h} = \frac{\pi}{2\lambda h}$$

in which λh is a non-dimensional parameter expressing the relative stiffness of the frame and infill, where

$$\lambda = 4\sqrt{\frac{E_i t \sin 2\theta}{4 E_c I h'}}$$

where E_i is the infill modulus, E_c is the frame modulus, I is the minimum moment of inertia of beams or columns and t is the wall thickness.

Failure occurs either by local crushing at the corner, diagonal splitting due to excessive compression, or shear failure. For typical frame geometry α equates to approximately $0.2h$, resulting in an effective width for the diagonal strut of $0.1w'$,

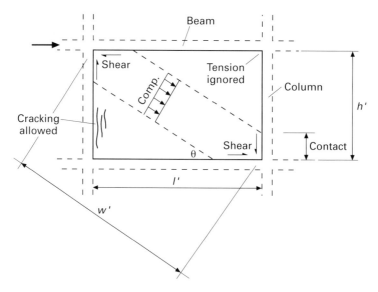

Fig. 9.21 Precast infill frame definitions and stresses

where $w' =$ diagonal length (corner to corner) of rectangular infill panel. The strength of the strut R_v is given by

$$R_v = 0.1kf_{cu}w't/\gamma_{m(conc)}$$

where $k = 0.3$ (stress factor for concrete in compression).

The horizontal resistance is given by $H_v = R_v\cos\theta$. Where infill slenderness ratio w'/t exceeds a certain value, usually about 12, the above equation is modified in accordance with wall buckling theory. Values of $H_v/l't$ are plotted graphically in Fig. 9.22[9.7] for a specified wall aspect ratio.

Hollow-Cored or Solid Cantilever Shear Walls

The design method for these walls is shown in Fig. 9.23. Hollow-cored wall elements are manufactured using grade C40 concrete and contain typically 0.2 per cent reinforcement in each face. Site-placed reinforcement includes starter bars cast in the foundation, to which additional bars are lapped to provide the holding down (tension) resistance. The spacing of the bars is typically between 100 mm and 250 mm depending on the strength requirements. *In situ* concrete (grade C40) is placed in the hollow cores to provide the compressive resistance.

The design is carried out for the ultimate limit state as follows. Referring to Fig. 9.24 the overturning moment M and axial compression N are calculated. The force in the reinforcing bars F_s and the concrete F_c are given by

$$F_s = nA_sf_y/\gamma_{m(steel)}$$

$$F_c = kf_{cu}bX/\gamma_{m(conc)}$$

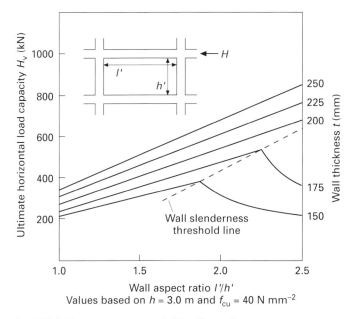

Fig. 9.22 Guide to precast concrete infill wall capacity

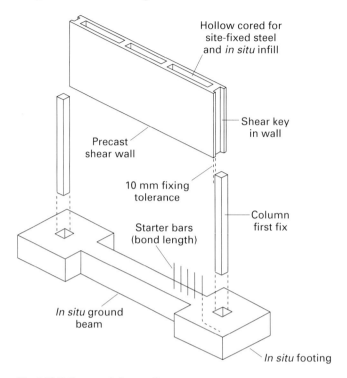

Fig. 9.23 Hollow-cored shear wall

where n is the number of bars in tension zone, i.e. in the zone $2(d - X)$ at a spacing s, where d is the distance to the centroid of the steel, and X is the depth to the neutral axis.

The combined stresses are predominantly flexural. The main difference between beam design and cantilever wall design is that the d and n will vary

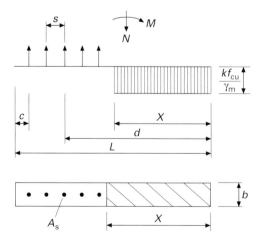

Fig. 9.24 Design of hollow-cored wall acting as cantilever beam

depending on the relative N and M magnitudes. Then

$$N = F_c - F_s = (kf_{cu}bX/\gamma_{m(conc)}) - (2(d-X)A_s f_y/s\gamma_{m(steel)})$$

$$M = kf_{cu}bXz/\gamma_{m(conc)} \quad \text{or}$$

$$= 2(d-X)A_s f_y z/s\gamma_{m(steel)}$$

depending on whether the steel or concrete is critical. From these equations X may be determined for given geometry, materials and loads. The minimum length of wall $L = d + (d-X) + \text{cover}$ to first bar (usually 50 mm) is compared with that available. If it is too great then the value of A_s/s must be increased.

Solid shear walls rely entirely on the connections in the horizontal and vertical planes. Horizontal joints may easily be formed using projecting dowels, and shear stress resistance equal to that of the precast concrete itself is possible. Vertical joints between adjacent precast wall panels are loaded mainly by shear forces, which are carried over either by using *in situ* concrete infill, embedded dowels or welded plates.

9.7   Horizontal Diaphragm Action

9.7.1   Introduction

There is no doubt that precast multistorey structures of up to about 50 m in height can be designed with economy, safety and excellent form. A vital aspect is the manner in which horizontal forces are transmitted through the components and their connections. There is also an increasing awareness towards the structural integrity of joints in prefabricated construction. The details used to achieve robustness in a structure have a significant effect on the structural mechanisms by which horizontal forces are distributed.

9.7.2   Horizontal Diaphragm Action in Precast Floors without Structural Screeds

In non-seismic zones horizontal forces in precast structures derive mainly from wind loading and temperature gradients. Horizontal loads are usually transmitted to shear walls or moment resisting frames through the roof and floors acting as *horizontal diaphragms*. Such action is analysed by considering the roof or floor as a deep horizontal beam, analogous to a plate girder or I-beam (Fig. 9.4). The critical situation is where the floors span parallel to the supporting shear walls. Equilibrating shear parallel to the supporting beam is carried by *shear friction* reinforcements which may be grouped at the columns or spaced equidistant between columns. The purpose of these bars is to transfer shear, caused by small lateral rigid body translations in the slab, into the chord elements. A tie force is thus generated in the reinforcement in the beam and/or *in situ* perimeter strip, where a short slot has been made in the tops of the hollow core to permit placement of the shear friction reinforcing bars (see Fig. 9.25).

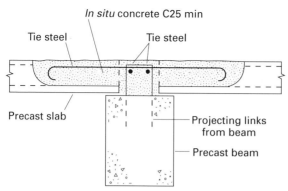

Fig. 9.25 Continuity of tie steel over internal beams acting as shear friction reinforcements in the floor diaphragm and continuity of internal tie steel

It is becoming increasingly popular to reinforce the *in situ* perimeter strip using seven-wire helical prestressing strand because the favourable mechanical properties of this type of strand and long lengths available on site make it an attractive alternative to high tensile bar. An anchorage length of at least 1200 mm for 12.5 mm diameter strand is used.

In this case an ultimate horizontal interface shear stress of about 0.23–0.27 N mm^{-2} is used in design (the actual value varies in different national codes). The depth of the unit can only be taken to the depth of the *in situ*/precast interface, i.e. the overall depth minus 30 mm in most types of slabs, which allows for differential camber and lack of adequate compaction at the bottom of the joint.

There have been extensive experimental studies of the shear transfer mechanism between hollow-cored slabs.[9.25–9.31] Wahid Omar[9.28] carried out full scale tests on 4 m long × 200 mm deep proprietary units and found ultimate load factors of at least 2.0 despite the presence of initial cracks up to 1 mm wide caused by the effects of grout shrinkage in the gap.

Shear strength and stiffness are provided by aggregate interlock, and ductility by dowel action of the reinforcing bars crossing the cracked interface. Cholewicki[9.32] has developed a mathematical model to predict the strength of the longitudinal joint based on shear friction and dowel action hypotheses.

9.7.3   Diaphragm Action in Composite Floors with Structural Screeds

Composite floor systems are designed on the basis that the precast flooring units provide restraint against lateral (in this case vertical) buckling in the relatively thin screed. The shear is carried entirely by the reinforced *in situ* concrete screed (the welded connections between double tee units are ignored). The design ultimate shear stress is at least 0.45 N mm^{-2} (for C25 concrete) and the effective depth of the screed is measured at the *crown* of prestressed flooring units.

Continuity of reinforcement in structural screeds is always extended to the shear walls or cores and it is safe to assume that the shear capacity of *in situ* diaphragms will not be the governing factor in the framing layout. Designers are careful in not allowing large voids near to external shear walls, and to ensure that if an external wall adjacent to a prominent staircase is used then a sufficient length of the floor plate is in physical contact with the wall.

9.8   Structural Stability

Stability of skeletal frames may be achieved in several ways as shown in Ref. 9.2, but in 90 per cent of cases the design is based on either:

1.   a *braced* structure (no-sway frame) (Fig. 9.26(a)); or
2.   an *unbraced* structure (sway frame) (Fig. 9.27).

As an alternative to every column resisting moments, a small number of large columns acting as *deep beams* may be used as wind posts at intervals along the structure. In all of these cases the stability design is an integral part of the component design.

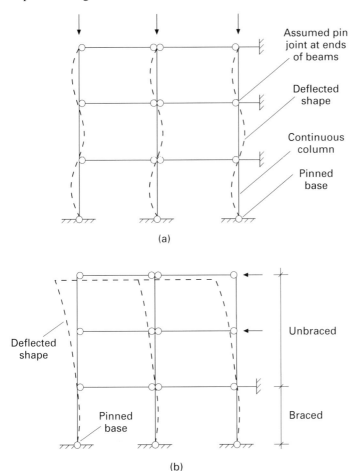

(a)

(b)

Fig. 9.26 (a) Braced structure, where bracing elements other than the skeletal components are used to provide stability; (b) partially braced structure

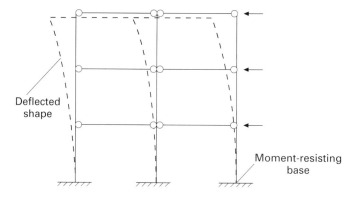

Fig. 9.27 Unbraced structure, which relies on column cantilever action for stability

A combination of the above is possible, which is known as a *partially braced* structure, Fig. 9.26(b). Here the structure is braced to a certain level (either down from the roof or, more likely, up from the foundation) and unbraced thereafter. Partial bracing does not refer to the nature of the joints, i.e. partial restraint. It is perfectly reasonable to brace the frame in one plane, and rely on column action in the other plane, particularly if the building is long and narrow in plan. However, it is generally accepted that the most economic solution, in terms of manufacture and frame erection, is to use the pin-jointed braced version, the main advantages being as follows.

- The number of precast elements is minimized, e.g. columns may be multistorey and beams need only be provided primarily in one direction.
- Pin-jointed connections are made on-site without the necessity of a seven-day (or longer) curing period required for *in situ* concrete joints.
- The frame is structurally stable floor by floor.
- Foundations are optimized because the column base is pin-jointed.

The stability of unbraced pin-jointed structures (Fig. 9.27) is provided entirely by columns designed as cantilevers for the full height of the structure. Partial restraints provided by moment-rotation or torsional stiffness in the beam–column connections, deep external spandrel panels or internal brick or block walls are ignored. New research is addressing these effects.[9.15,9.33]

The maximum overturning moment in each column is $\sum H_i h_i$, where H_i is the floor diaphragm reaction at each column, and h_i is the effective height from a point 50 mm below the top of the foundation to the centroid of the floor plate i. In most instances the columns will be equally loaded horizontally. The overturning moment is additive to the frame moments derived under column design. There is no moment distribution into the beams because the connections are pinned.

The maximum number of storeys for an unbraced pin-jointed structure is about three, giving a maximum height of about 10 m to 12 m. Even at this height architectural restrictions on the sizes of columns and the magnitude of the moment-restraint required at the foundation are likely to be prohibitive.

Braced structures offer the best solution to stability in multistorey construction, irrespective of the number of storeys. Connection details and foundation design and construction are greatly simplified. Precast concrete wall units are inexpensive, have large in-plane stiffness and strength, are easy to erect and may be integrated with the structure using either the infill wall or cantilever wall (or box). Other methods of bracing are infill brick or block walls, and steel cross-bracing. The distribution of horizontal loading between shear walls or frames is dependent upon the following.

1. *In-plane deflection response.* This is predominantly a flexural deflection in cantilever walls, a combined flexural and shear deflection in infill walls, and a truss deflection in steel cross-bracing.
2. *Position.* The structure must be *balanced* by disposing the walls according to their stiffness and in such a manner that the centre of pressure of horizontal loading lies between at least two of the 'larger' walls. Torsional effects, resulting from eccentricities are statically determinate and modifications to the load distribution satisfy equilibrium in the direction of loading.[9.34]
3. *Expansion joints in the floor diaphragm.* In general, structures exceeding 80 m in plan dimension are usually isolated.

The moments in the columns are small, but the axial forces have to be considered, particularly in the lightly gravity-loaded cases where walls are positioned between gable columns and uplift may occur. On the other hand, large compressive forces may result in the columns adjacent to infill walls requiring a greater section (or reinforcement) than the remainder. Columns adjacent to cantilever shear walls are assumed not to carry additional axial forces due to overturning moments, despite the presence of a continuous vertical shear key between the column and wall.

9.9   Robustness and Resistance Against Progressive Collapse

In the event of accidental damage or abnormal loading, an alternative load path must be created to guard against a progressive collapse. It is usual that tie forces between the precast components will be mobilized.[9.35,9.36] Three alternative methods are possible:

- use of ties;
- protected members;
- alternative load paths.

The last two cases are classified as 'direct' methods because an appreciation of the severity and possible location of any accidental damage is known or assumed. In the first case a notional tie force is calculated and continuity reinforcement is provided between slabs over beams and walls, and for beams over columns. The ties must run continuously, in both the physical and structural sense, around the full perimeter and along all internal beam lines of the building and be capable of protecting *all* precast components from failing if an adjacent member itself fails.[9.37]

Internal ties parallel with the span of the flooring are either distributed evenly using short lengths of tie steel anchored by bond into the opened cores of the hollow-cored floor units, or grouped in full-depth *in situ* strips at positions coincident with columns. Internal and peripheral ties parallel with the span of beams are grouped at the beams. To prevent failure by vertical splitting, these ties pass beneath the floor ties and are placed inside projecting loops, hooks or similar projecting beam reinforcement, Figs 9.25 and 9.28.

The form of the lateral tie steel may be either a deformed high tensile bar, lapped at positions away from the connections, or a continuous piece of seven-wire helical strand (typically 12.5 mm dia.) placed into a narrow (typically 50–80 mm wide) *in situ* concrete strip between the ends of the flooring units and supporting beams. This reinforcement, which is also provided for floor diaphragm purposes, provides the necessary tie forces required to allow the precast elements to move relative to one another but to absorb the energy created under accidental forces. Test results[9.28] have shown that the full anchorage bond length for 12.5 mm diameter strand should be 1.20 m. Tests by precast manufacturers have shown that bend radii of 400 mm are possible without detriment to the anchorage characteristics of the strand at the corners.

A continuous vertical tie is provided in all precast frames independent of height. The physical connection made at column splices ensures that an adequate vertical tie force is generated. The horizontal joints between loadbearing wall panels are specifically checked in this respect because the normal gravity loading condition eliminates the necessity for tensile capacity.

Fig. 9.28 Continuity ties at external columns

9.10   Future Developments in Precast Structures

The construction industry requires a multiple choice in the selection of building components. In the next few years it is likely that the increasing demand on the performance of these components will overtake existing technology and designers will be left with no other choice than to extrapolate existing knowledge to meet this demand. The call for higher specifications for commercial, industrial, civic and domestic buildings has led to a major rethink in construction strategy. The precast concrete industry is ideally placed to accommodate these higher demands because it is widely appreciated that the degree of prefabrication, using skilled labour in quality controlled conditions, is set to increase dramatically.

This chapter has shown that the principle of prefabricating concrete is most profitable if the structural system is chosen such that the components can be optimally designed, produced and assembled. This is of special significance in precast prestressed floors, since about 60 to 70 per cent of the total building material is used in floors. To maintain competitiveness, it is appropriate that the precast concrete industry is engaged in R & D to extend the application of recently available technologies to the development of an enhanced structural system.

The main industrial objective during the next 10 years is the improvement in the structural integrity of the buildings, leading to a higher degree of safety and reliability through the development of better design and manufacturing techniques. This includes the recuperation of waste products, specifically recycled concrete crushed for use as a partial or total replacement for natural aggregates and by the use of by-product material, such as silica fume and nodulized fly-ash from blast furnace waste. New design techniques are being developed which will lead to lighter construction by using high strength concrete, partial prestressing and post-tensioning techniques and the optimization of the cross-sectional profile. Studies are being carried out to generate data on the functional requirements of high specification buildings in the modern world, deterioration in structural performance due to vibrations, fatigue etc., the real behaviour of connections and associated stability problems, and on horizontal diaphragm action under seismic, cyclic and ultimate load conditions. The success and implementation of this work will vary in different countries, but the net result is intended to make buildings:

- cheaper, through faster, safer and easier construction, and usage of less material in the precast structural framework;
- more reliable and durable;
- more architecturally and structurally versatile.

It will allow:

- longer spans, or greater load-carrying capacity;
- changes in the functions of building;
- greater architectural freedom;
- fewer geometric restrictions on building form.

The requirement for off-site fabrication will continue to increase because the rapid growth in management contracting, with its desire for reduced on-site occupancy and high quality workmanship, favours controlled fabrication methods. The construction industry must be made aware of the maximum benefits of precast concrete by widespread education and training programmes. Precast concrete must be seen as a philosophy and not as a building product.

References

9.1   Fédération Internationale de la Précontrainte 1986 *Design of Multi-Storey Precast Concrete Structures* FIP Commission on Precasting, Thomas Telford, London 27pp

9.2   Fédération Internationale de la Précontrainte 1994 *Planning and Design of Precast Concrete Structures* FIP Commission on Prefabrication, SEKO, Institution of Structural Engineers, London, 138 p

9.3   Bljuger F 1988 *Design of Precast Concrete Structures* Ellis Horwood, Chichester

9.4   Huyghe G, Bruggeling A S G 1991 *Prefabrication with Concrete*, Balkema, Rotterdam, 380 p

9.5   Sheppard D A, Phillips W R 1989 *Plant-Cast and Prestressed Concrete* 3rd edn, McGraw-Hill 791pp

9.6   Elliott K S 1996 *Multi-storey Precast Concrete Framed Structures* Blackwell Scientific, Oxford

9.7   Elliott K S, Tovey A 1992 *Precast Concrete Framed Buildings – A Design Guide* British Cement Association, Wexham Springs, 88 p

9.8   ISE 1978 *Structural Joints in Precast Concrete* Institution of Structural Engineers, London, 56 p

9.9   PCI 1985 *PCI Design Handbook* Prestressed Concrete Institute, Chicago, 510 pp

9.10   Fédération Internationale de la Précontrainte 1988 *Precast Prestressed Hollow Cored Floors* FIP Commission on Precasting, Thomas Telford, London, 31 pp

9.11   Walraven J G, Mercx W P M 1983 The bearing capacity of prestressed hollow-core slabs. *Heron* **28**(3): 1–46

9.12   Elliott K S, Davies G, Adlparvar R M 1993 Torsional behaviour of joints and members in precast concrete structures. *Magazine of Concrete Research* **164**: 157–168

9.13   Martin L D, Korkosz W J 1982 *Connections for Precast Prestressed Concrete Buildings, including Earthquake Resistance* The Consulting Engineers Group, Inc, Illinois 60025, 288 p

9.14   PCI – *PCI Manual on Design of Connections for Precast Prestressed Concrete*, Prestressed Concrete Institute, Chicago (see also Ref. 9.9)

9.15   Elliott K S, Davies G, Mahdi A A 1992 Semi-rigid joint behaviour on columns in precast concrete buildings. *COST C1 Seminar* ENSAIS, Strasbourg, 282–295

9.16   Mitchell D, Marcakis K 1980 Precast concrete connections with embedded steel inserts. *PCI Journal* July–Aug: 88–116

9.17   Holmes M, Posner C D 1971 *The Connection of Precast Concrete Structural Members* CIRIA Report 28, London, 47 p

9.18   Cook W D, Mitchell D 1988 Studies of disturbed regions near discontinuities in reinforced concrete members. *ACI Structural Journal* March: 206–216

9.19   Dougill J W, Kuttab A 1988 Grouted and dowelled jointed precast concrete columns: behaviour in combined bending and compression. *Magazine of Concrete Research* **40**(144): 131–142

9.20   Korolev L V, Korolev H V 1962 Joint between prefabricated reinforced column and foundation. *Promyshlennoe Sroitel'stvo* **16**(9)

9.21   Stafford-Smith B, Carter C 1969 A Method of Analysis for Infill Frames. *Proceedings of the Institution of Civil Engineers* **44** (Sept): Paper 7218, 31–48

9.22   Mainstone R J 1972 *On The Stiffnesses and Strengths of Infill Frames* Building Research Establishment Paper CP2/72, Feb, 35 p

9.23   Wood R H 1978 Plasticity, composite action and collapse design of unreinforced shear wall panels in frames. *Proceedings of the Institution of Civil Engineers* **65** (June): Paper 8110, 381–411

9.24   Kwan K-H, Liaum T-C 1982 Non-linear analysis of multi-storey infilled frames. *Proceedings of the Institution of Civil Engineers* **73** (June): Paper 8577, pp. 441–454 (see also Plastic theory of infilled frames with finite interface shear strength. *Proceedings of the Institution of Civil Engineers* **75** (Dec 1983): Paper 8718, 707–723)

9.25   Moustafa S E 1981 Effectiveness of shear-friction reinforcement in shear diaphragm capacity of hollow-core slabs. *Journal of the Prestressed Concrete Institute* **26**(1) 118–132

9.26   Svensson S 1985 *Diaphragm Action in Precast Hollow-Core Floors, Description of a Pilot Test Series,* The Nordic Concrete Federation, Nordic Concrete Research, Publication No 4

9.27   Svensson S, Engstrom B, Cederwall K 1986 Diaphragm action of precast floors with grouted joints. *Nordisk Betong* (Feb): 123–128

9.28   Omar W 1990 *Diaphragm Action in Precast Concrete Floor Construction* PhD Thesis, University of Nottingham

9.29   Davies G, Elliott K S, Omar W 1990 Horizontal diaphragm action in precast concrete floors. *The Structural Engineer* **68**(2): 25–33

9.30   Elliott K S, Davies G, Omar W 1992 Experimental and theoretical investigation of precast hollow cored slabs used as horizontal floor diaphragms. *The Structural Engineer* **70**(10): 175–187

9.31   Menegotto M 1989 Seismic resistant extruded hollow core slabs. *International Symposium on Noteworthy Developments in Prestressed and Precast Concrete* Singapore, November

9.32   Cholewicki A 1991 Shear transfer in longitudinal joints of hollow core slabs. *Concrete Precasting Plant and Technology* No 4: 58–67

9.33   CEC 1991–1995 *COST C1: The Control of Semi Rigid Behaviour of Civil Engineering Structural Connections* Commission of the European Communities, Rue de la Loi 200, B-1049 Brussels, Belgium

9.34   Hogeslag A J 1990 Stability of precast concrete structures. *Prefabrication of Concrete Structures* International Seminar, October, Delft University of Technology, Delft University Press: 29–40

9.35   Department of Environment and CIRIA 1973 *The Stability of Precast Concrete Structures* Seminar, London, March

9.36   Engstrom B 1992 *Ductility of Tie Connections in Precast Structures* PhD Thesis, Chalmers University of Technology, Goteborg, Sweden

9.37   Schultz D, Burnett E, Fintel M 1977 *A Design Approach to General Structural Integrity – Design and Construction of Large Panel Concrete Structures* Report 4, Portland Cement Association, Skokie, USA

10 Movement Joints: a Necessary Evil, or Avoidable?

Jürgen Ruth

Movement joints are traditionally used in concrete buildings to reduce or eliminate the internal restraint forces which develop as the result of phenomena such as temperature change, foundation movement and concrete shrinkage. The effect of a joint is to concentrate large deformations in a localized region (i.e. the joint) and hence prevent the severe cracking which would otherwise occur.

There is considerable evidence from existing buildings to suggest that movement joints often do not act in the manner intended by the designer. They can in fact have a very adverse effect on the performance of concrete structures.

This chapter explores methods for controlling differential movements and restraint forces in large concrete buildings, with a view to eliminating the need for joints. For those situations where it is not possible to eliminate joints altogether, the problems of locating and detailing joints are also discussed briefly.

10.1   Introduction

The structural behaviour of a concrete building is influenced not only by the applied loads but also by various load-independent effects, such as temperature variations, differential foundation movements, and concrete shrinkage. The way a building responds to load-independent effects depends on the relative stiffnesses of its structural components.

Figure 10.1 shows how bending occurs in the raft footing of a high-rise building which consists of a high tower and a much lower surrounding podium, all sitting on a soft foundation material.

In Fig. 10.2 temperature effects are illustrated for a skeletal structure (a) which is stiffened by end cores.[10.1] The effects of heating and cooling in the slab floors, e.g. as the result of air-conditioning, are shown in (b) and (c). The results of solar radiation on the roof slab (d) and fire in several intermediate floors (e) are also shown. Effects comparable to those shown in Fig. 10.2 can also result from drying shrinkage of the concrete.

Large restraint forces can develop within a building structure as the result of load-independent effects, and they can result in significant differential movements and severe cracking. Joints are traditionally used to reduce or eliminate the restraint forces. They concentrate deformations in small predefined regions (the joints) and thereby prevent severe cracking from occurring over extended regions. For example, movement joints can be used to accommodate the relative vertical displacements shown in Fig. 10.1 and the differential horizontal expansions illustrated in Fig. 10.2. Questions concerning detailing and locating movement joints in concrete buildings are taken up in Sections 10.3 and 10.4.

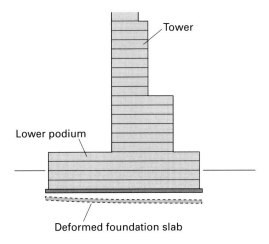

Fig. 10.1 Bending of raft footing

The damage observed in buildings in recent decades[10.2] suggests that although the basic concept of joints may be correct, the intended result is often not achieved in practice. While the reason for this may lie partly in an incorrect choice of joint details, high precision in construction is necessary if the joint is to work successfully, and in many cases this apparently cannot be achieved on site because of extreme pressures on time, which derive from the prevailing 'time is money' philosophy of construction.

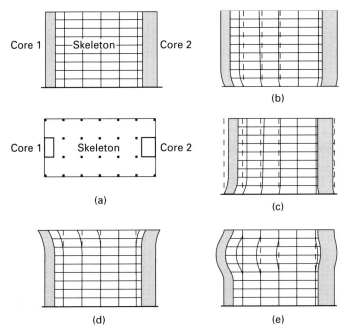

Fig. 10.2 Temperature effects in a building frame

Even apart from construction problems, there are other significant disadvantages to the use of joints. These include:

- increased construction time;
- increased construction costs;
- disturbance of the visual homogeneity of external walls and internal partitions and slabs;
- expensive maintenance;
- decreased stiffness, possibly leading to vibration susceptibility;
- increased permeability for gases and liquids;
- various adverse effects in the context of building physics.

As a consequence, designers should always investigate the possibility of eliminating joints in buildings,[10.3] or, if this is not feasible, of using as few joints as possible. In undertaking such an investigation, existing tables and rules of thumb for estimating the necessary numbers of joints[10.4] should be avoided.

10.2   Avoiding Joints

Joints can be avoided if, firstly, the restraint forces and deformations can be kept to acceptably low levels, and, secondly, the building is designed to withstand the deformations and forces which occur. These two basic aspects of avoiding joints in buildings will be discussed in turn.

10.2.1   Limiting Restraint Forces and Deformations

Methods are considered here for limiting the restraint forces and deformations to a level which makes joints unnecessary. Attention is restricted to three sources of restraint forces and deformations: differential settlement under load, temperature effects and shrinkage.

Differential Settlement under Load
It may be possible to reduce differential settlement by adjusting the depth of the foundation, and thereby the modulus of the subgrade reaction, in accordance with the varying height of the building. This is shown in Fig. 10.3. Heavily loaded tower regions of the building can thus be induced to settle to the same extent as other less heavily loaded regions. Ideally, the footing slab then experiences only the local stresses due to the column loads, and the effects of transitory live loads and horizontal loads.

Other possibilities for reducing differential settlement include:

- adjusting the soil properties by soil replacement, grouting or compaction;
- introducing stiffening elements such as piles and diaphragm walls.

Another relatively new method is to provide piles to carry the loads transferred to the foundation slab. This is the so-called combined pile–slab foundation[10.5] and is shown in Fig. 10.4. The pile pattern can also be adapted to increased loads in the tower area by a closer spacing.

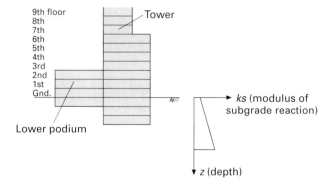

Fig. 10.3 Reduction of differential settlement in a building of varying height

In especially critical situations where the footings occur on soft soil with the potential for large differential settlement, an active solution can be employed. Compensation in regions of high (or low) settlement may be achieved by reducing (or increasing) the local vertical deformations of the foundation slab. Figure 10.5 shows a construction technique for compensating for low settlement by using water-filled Freyssinet-type flat jacks.[10.6] The inclination which would otherwise develop in the building is eliminated by elevating portion of the foundation slab. In contrast, ballasting can be used to produce a local increase in settlement.[10.7] An alternative to ballasting is to apply vertical forces on the foundation by prestressing anchored vertical tension members. Professor Leonhardt has proposed this approach as a means of reducing the inclination of the 'Leaning Tower' of Pisa.

Although techniques like the ones described can be very effective, they are also expensive, and they prolong the construction time. The economic efficiency of such measures therefore has to be considered carefully. On the other hand, if unfavourable soil properties and poor quality endanger the structure there may be no alternative to costly steps such as these.

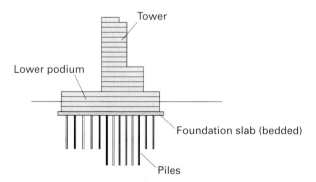

Fig. 10.4 Placement of piles beneath a raft foundation

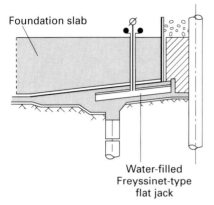

Foundation slab

Water-filled
Freyssinet-type
flat jack

Fig. 10.5 Compensation for low settlement using water-filled Freyssinet-type flat jacks[10.6]

Deformations Caused by Temperature and Shrinkage

Various methods can be employed to reduce the adverse effects of temperature and shrinkage. For example, the influence of shrinkage can be reduced either by limiting the susceptibility of the structure to shrinkage deformations, or by delaying the completion of construction in shrinkage-sensitive regions until the tensile strength of the young concrete reaches a sufficiently high level to carry any tensile stresses which develop.

Improved tensile strength can be achieved through the choice of a suitable concrete mix with a low cement content and a low water/cement ratio. During curing, the concrete can be kept moist and drying can be prevented by the use of covers. Temporary thermal insulation can also be applied at the time of placement. Placement of the concrete in a foundation slab can be sequenced in such a way that shrinkage stresses are minimized, for example by the use of a checker-board pattern in which only alternate squares are concreted at the same time.

In order to reduce the effects of temperature it makes sense to start concreting at a medium temperature and hence limit the maximum temperature changes which will be experienced by the concrete. Of course, thermal insulation can also be used to reduce fluctuations in temperature.[10.8]

The methods described here can only reduce, but not eliminate, the effects of temperature changes and shrinkage of the concrete. It is therefore important to consider carefully the layout of the structural system already in the preliminary stages of design. Figure 10.6 shows how the location of a core in a structure

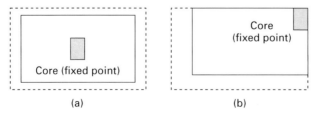

(a)   (b)

Fig. 10.6 (a) Favourable and (b) unfavourable core locations in a building

greatly influences the relative deformations which occur in a slab and hence in the attached columns, as the result of changes in temperature.

Figure 10.7 shows further considerations for locating stiffening elements.[10.9] Generally, the placement of cores or walls in extreme and remote locations can result in large stresses both in these elements and in the connecting slabs, and should be avoided.

Sometimes a favourable placement and a partial connection of columns can reduce the build-up of restraint forces in the structural system. Partially hinged columns are clearly far less sensitive to temperature changes, and may be used in certain circumstances, provided overall stability is not compromised.[10.10]

Special solutions incorporating adjustable bearings or viscous bearings may also reduce the stresses due to unavoidable strains in structural elements. During the construction of the 'Katharinenhospital', adjustable bearings were used to connect a square slab with four cores at the corners to reduce the initially large stresses caused by shrinkage of the concrete.[10.11]

10.2.2   Design Calculations for Jointless Construction

The methods discussed above can alleviate, but never eliminate completely, the effects of differential settlement, restrained shrinkage and temperature. The stresses which remain have to be evaluated carefully, so that suitable reinforcement can be employed to control the cracking of the concrete.

The building shown in Fig. 10.8 is just one example from many of the successful designs of a sophisticated large building without any joints.[10.12] Many more buildings constructed without joints have been described by Falkner.[10.13,10.14]

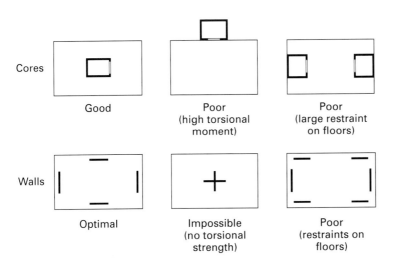

Fig. 10.7 Locations for stiffening elements[10.9]

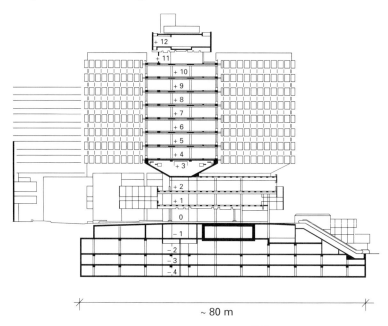

Fig. 10.8 Large building without joints (GENO building)[10.12]

Careful adherence to a number of design principles and calculation procedures, as discussed below, ensures that all design requirements are met over a long period of service.

Non-Linear Behaviour

The internal restraint force induced by an imposed deformation depends very much on the stiffnesses of the restrained member. In the case of reinforced concrete, the structural behaviour is non-linear and this must be considered carefully. Non-linearity is actually an advantage, because the reduction in stiffnesses with increasing load, due to progressive cracking, leads to smaller restraint forces. The analysis can be undertaken using modern computer programs; alternatively, design charts are available which are based on theoretical analysis combined with test data. Some sample charts are considered here.

Column Restraints. Figure 10.9 shows test results for the relative displacement of the top of a cantilevered column as affected by axial force and the arrangement of the stirrup reinforcement.[10.1] The straight line represents a lower limit to the deformations as determined from the tests and corresponds well with theoretical calculations reported by Hock *et al.*[10.1] The figure can be used in the design of columns subjected to imposed deformations. For columns fixed at both ends against rotation, the allowable relative lateral displacement appears to be only half as large as for a cantilevered column.[10.15]

Another useful effect is shown in Fig. 10.10. Rotations of the column head, due to elongation of the slab floor, are somewhat reduced by the rotations which also

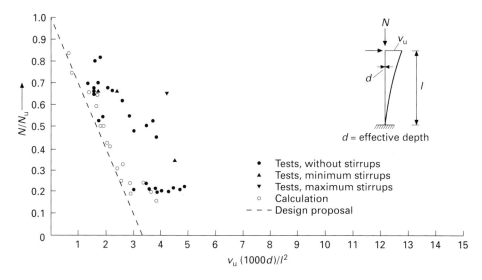

Fig. 10.9 Displacement v_u at the top of a cantilevered column[10.1]

occur in the slab floor.[10.1] This figure also shows the relevant substitute column systems which can be used for analysis.

The diagram in Fig. 10.11 can be used to estimate the disadvantageous effect of a nodal rotation on the allowable horizontal displacement v_u at the top of a column. The nodal rotation is caused by the end moment M_r induced by loading

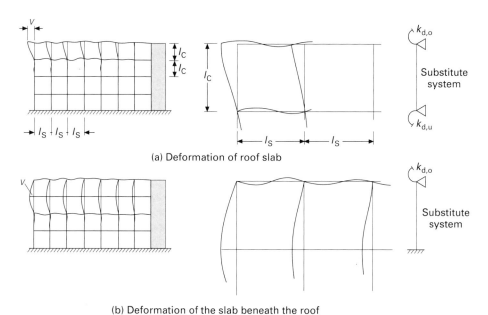

Fig. 10.10 Column deformations due to slab elongation[10.1]

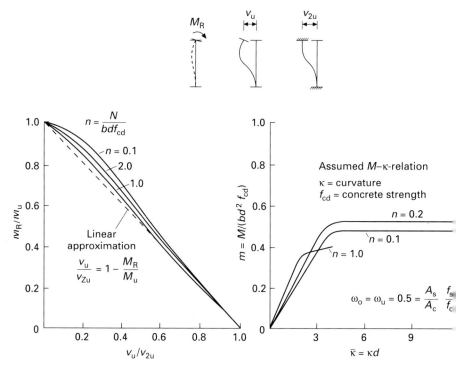

Fig. 10.11 Reduction in allowable displacement at top of column[10.1]

on the adjacent horizontal floor member.[10.1] The allowable displacement v_{2u} is for a column which is rigidly fixed against rotation at each end.

The two diagrams in Fig. 10.12 from Ref. 10.1 show how increased lateral displacement v at the top of a doubly reinforced column requires an increase in the strain ε_1 in the tensile column face. The strain ε_1 is related to the crack width w and therefore has to be limited. The increase in ε_1 (shown on the horizontal axis) with v (vertical axis) is initially linear, but there is a sharp increase in the rate at higher levels when cracking occurs. The effect of axial force N is to reduce ε_1.

Slab Restraints. The adverse effect of axial restraint on a slab in tension is an increase in crack width. Restraints causing axial compression are usually favourable, although the concrete compressive stresses should be checked. When tensile axial restraint occurs, it is important to ensure that the cracks are evenly distributed. Wide cracks are likely to occur in regions where the section of the slab is weakened by recesses, stairwells, local reductions in thickness, and at special connections with other structural members. In such regions the reinforcement should exceed the minimum requirements mentioned below. The stiffness of a reduced region should be the same as for normal regions. As an alternative, it is possible to concentrate all of the elongation Δl in a small

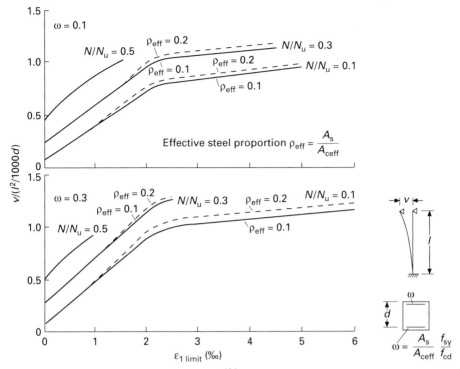

Fig. 10.12 Allowable end displacement *v* of column[10.1]

weakened region by suitable arrangement of the reinforcement to minimize crack widths elsewhere.

It is often forgotten that restraints in complex disturbance D-regions of a structural system have to be checked carefully to keep cracks small and harmless. Strut-and-tie modelling, as discussed by Reineck in Chapter 5 of this book, provides a convenient means for identifying and designing for the flow of forces in such regions.

Restraints in Stiffening Cores and Walls. Any elongation or shortening of a slab floor in a building induces forces in the cores or stiffening walls, and the magnitude of such forces depends on the relative stiffnesses of the participating elements. A non-linear frame analysis using an appropriate computer program is needed if the favourable effect of the flexibility of the stiffening elements is to be taken into account.

If restraint forces are to be calculated for catastrophic loading conditions, such as fire, it is reasonable to assume that the relevant stiffnesses are sharply reduced because of the severe cracking which results from yielding and possibly fracture of the reinforcement.[10.15] The loadbearing capacity of the stiffening elements under vertical loading is nevertheless not reduced.

Reinforcement Requirements for Crack Control

Structural members always have to be provided with sufficient reinforcement to prevent the crack widths from exceeding the permissible limits, which depend mainly on the environmental conditions and the specific utilization of the structure. For example, the standards of impermeability of an external basement wall of an archive building will be higher than for an underground parking station.

The required reinforcement level depends on the dimensions of the member, the quality of the concrete and reinforcing steel, the layout of the reinforcement and the applied loads and restraints.

The calculation of the reinforcement required for crack control should be based on a model of the bond between the reinforcement and the concrete which takes account of the main influences.

Schober[10.16] uses strut-and-tie models to develop a design diagram for a tension bar (see Fig. 10.13). This is a key element in treating the internal flow of forces in any reinforced concrete structure. From the non-linear relation between axial force and elongation under conditions of either applied load or restraint (Fig. 10.14), the required design diagrams can be derived analytically, without recourse to empirical correction factors.

The parameters which define the tension stiffening effect $\Delta\varepsilon_s$ (see Fig. 10.14), are primarily the concrete tensile strength and the bond quality between steel and concrete. These have an important influence on the required amount of reinforcement. Figures 10.15 and 10.16 contain design diagrams by Schober for cases of applied load and restraint, respectively, and allow the immediate determination of the required percentage of reinforcement ρ_{eff}. The value depends

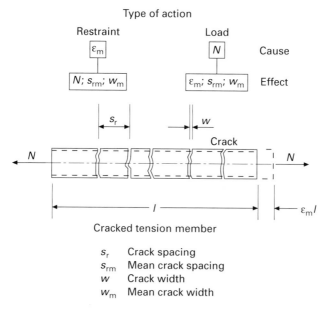

Crack spacing s_r Crack spacing
s_{rm} Mean crack spacing
w Crack width
w_m Mean crack width

Fig. 10.13 Tension member, loaded or restrained

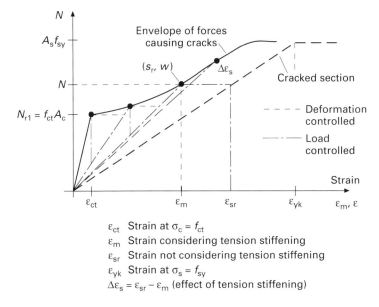

ε_{ct} Strain at $\sigma_c = f_{ct}$
ε_m Strain considering tension stiffening
ε_{sr} Strain not considering tension stiffening
ε_{yk} Strain at $\sigma_s = f_{sy}$
$\Delta\varepsilon_s = \varepsilon_{sr} - \varepsilon_m$ (effect of tension stiffening)

Fig. 10.14 Axial force N versus strain ε_m, load and deformation control

on the bar diameter d_s, the ratio of permissible crack width to bar diameter, w_m/d_s, and either the imposed restraint strain ε_{tm}, or the steel stress σ_s due to the applied load.

An interesting variation on the above design diagram is shown in Fig. 10.17, where the expansion due to restraint ε_{tm} directly influences the percentage of reinforcement ρ_{eff} required to guarantee a specific maximum crack width w_m. Of special note is the case of a small crack width, $w_m = 0.1$ mm, which could be specified to ensure impermeability of an external face of a waterproof concrete basement. With a bar diameter of $d_s = 10$ mm and a ratio of $w_m/d_s = 0.01$, a large amount of reinforcement, about $\rho_{eff} = 1.06$ per cent, is required in order to achieve small strains of $\varepsilon_{tm} = 0.1$.

If load and restraint act in sequence, the reinforcement may be determined in the same way as for restraint only. Should the stresses due to load require a higher percentage of reinforcement the increase in steel stresses due to restraint can be estimated with $\Delta\sigma_s = \varepsilon_{tm}E_s$. The crack width can be limited using the diagram shown in Fig. 10.16 and the equation: $\sigma_s = \sigma_{sD} + \Delta\sigma_s$.

Here the steel stress σ_{sD} is due to the frequently acting part of the load, and reaches at most 70 per cent of the permissible working load in large concrete buildings.

Assessment of Restraint ε_{tm} and Influence of Creep

The magnitude of the imposed deformation influences the required amount of reinforcement, as Fig. 10.17 demonstrates. The assessment of the actual elongation due to a restraint is, however, often not so easy. Firstly, the potential temperature change or shrinkage strain is not precisely known, but must often be

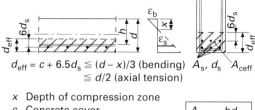

$$d_{eff} = c + 6.5d_s \leqq (d - x)/3 \text{ (bending)} \qquad A_s, d_s \quad A_{ceff}$$
$$\leqq d/2 \text{ (axial tension)}$$

x Depth of compression zone
c Concrete cover
d_s Bar diameter

$$\boxed{A_{ceff} = bd_{eff}}$$

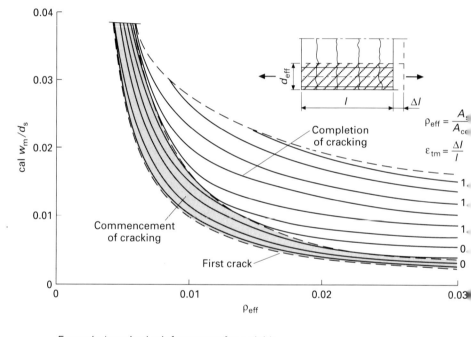

Completion of cracking

$$\rho_{eff} = \frac{A_s}{A_{ce}}$$

$$\varepsilon_{tm} = \frac{\Delta l}{l}$$

Commencement of cracking

First crack

Fig. 10.15 Reinforcement for crack control, condition of restraint[10.16]

Example (required reinforcement for a slab):

cal w_m = 0.20 mm (w_{max} = 1.7 w_m = 0.34 mm); c = 25 mm; d = 300 mm

d_s = 10 mm, d_{eff} = 25 + (6.5 × 10) = 90 mm < d/2 = 300/2 = 150 mm (axial tension)

$$\left. \begin{array}{l} \dfrac{\text{cal } w_m}{d_s} = \dfrac{0.20}{10} = 0.020 \\[2mm] \varepsilon_{tm} = 0.20\text{‰} \end{array} \right\} \quad \rho_{eff} = 0.0068$$

req. a_s = 0.0068 × 90 × 1000 = 612 mm^2 m^{-1} = 6.1 cm^2 m^{-1}

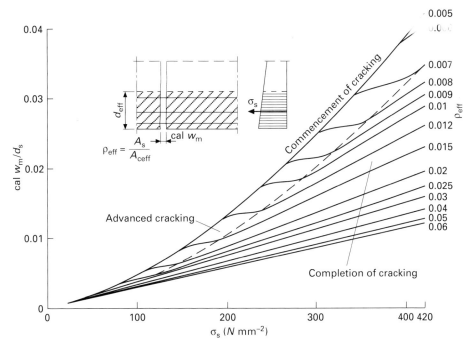

Fig. 10.16 Reinforcement for crack control, condition of load[10.16]

assessed with engineering judgement. Secondly, the elongation of a cracked structural concrete member under load may already include some component of the axial restraint. Finally, the flexibility of the stiffening members providing restraint to the member can often only be assessed, unless a fully non-linear frame analysis for the whole structural system is carried out, as previously discussed.

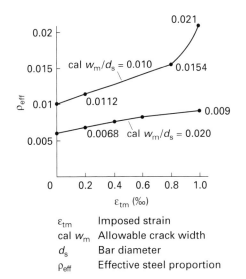

ε_{tm}	Imposed strain
cal w_m	Allowable crack width
d_s	Bar diameter
ρ_{eff}	Effective steel proportion

Fig. 10.17 Required reinforcement

The proper consideration of the favourable effect of concrete creep may also result in a considerable reduction in stress, especially when the imposed strain increases slowly, e.g. due to shrinkage or progressive settlement. Since the concrete is creeping simultaneously, the stresses are somewhat reduced.

10.3   Locating and Constructing Movement Joints

If the internal restraint forces are large enough to compromise the load capacity or serviceability of a building, even after all realistic design countermeasures have been considered, movement joints will have to be introduced in order to cater for the necessary movements and to reduce the restraint forces to acceptable levels. The number of joints should nevertheless be kept to a minimum, and the joint details should be as simple as possible. A number of characteristic joint designs will be discussed, with variations in detailing.

10.3.1   Settlement Joints for Differential Vertical Displacements

In Fig. 10.18 details are given of a typical settlement joint in a slab footing.[10.17] The joint has been placed between the tower region, where large settlements occur, and the lower podium where the settlement is considerably less.

Details of a possible joint in another building are shown in Fig. 10.19.[10.18] A hinged cover plate is used to accommodate the potentially large differential settlement between the podium and tower regions.

10.3.2   Movement Joints for Differential Horizontal Displacements

If the decision has been made to include joints to accommodate differential horizontal movement, possible locations need to be considered in order to find the optimum. Figure 10.20 shows just three possible arrangements for the separation of two cores. In addition to complete separation, solutions with fewer degrees of freedom of movement are possible which aim to reduce the restraint forces to acceptable levels, yet maintain some structural rigidity.

Reduction in structural stability and stiffness can be minimized by providing expansion joints only in critical regions of high stiffness where the maximum restraint forces develop. Another possibility is to restrict movement only in one direction. Figure 10.21 gives an example in which the joints are restricted to the lower floors where the building is at its stiffest, and the potential effect of the joints is maximum.[10.19]

Many innovative jointing arrangements have been developed to solve specific design problems in large concrete buildings, and are reported in the literature. For example, special heavy sliding bridge bearings were used to separate the lower-most floor of a building from the underlying foundation.[10.19]

Joints may be unavoidable, even in prefabricated construction. In Fig. 10.22 details are shown of a joint which transmits shear forces while preventing stresses developing from temperature deformations. The joint extends through the full width of this long building without drastically reducing its resistance to horizontal

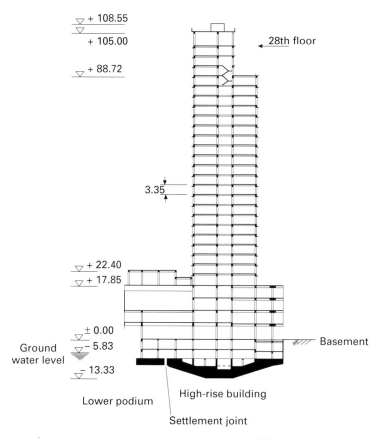

Fig. 10.18 Settlement joint (differential vertical movement)[10.17]

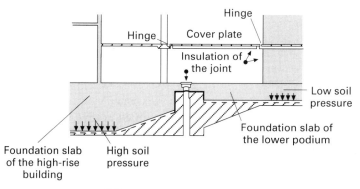

Fig. 10.19 Detail of joint[10.18]

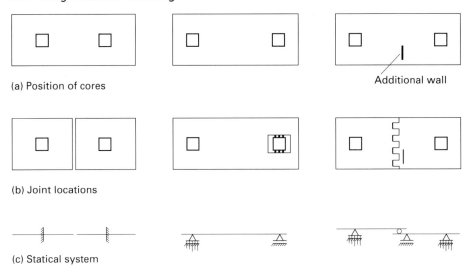

(a) Position of cores

Additional wall

(b) Joint locations

(c) Statical system

Fig. 10.20 Locations for horizontal movement joint

forces.[10.20] The building in this case is the Züblin Building, an aesthetically convincing large building, constructed from precast elements.

10.4   Joint Detailing

Since joint detailing is well covered both in the technical literature and in commercial brochures, this discussion is brief. Several examples are given which show the importance of adapting the joint details to the requirements of good structural performance under load, and durability.

In designing and detailing a joint, the first question to be considered is whether complete physical separation is required, or partial separation with the transfer of some force components. Joints with complete separation are the least complicated and therefore most reliable and easiest to design and construct.

For the complete separation of a slab system, the joint may be placed either in the web or between webs (Fig. 10.23). Special details can be chosen to allow for the smooth transition of the slab surface across such joints.

Figure 10.24 shows an arrangement for a joint which involves relative horizontal sliding with force transfer by the introduction of special pads, without any adverse effects on appearance or waterproofing. Various details have been worked out to transfer forces both transverse and parallel to the joint. Strut-and-tie models provide a convenient basis for detailing the reinforcement in the regions immediately adjacent to the joint.[10.21,10.22]

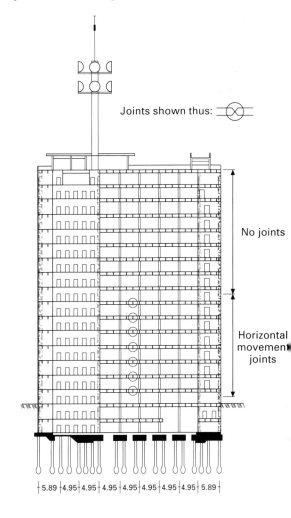

Joints shown thus: ─⊗─

No joints

Horizontal movement joints

|5.89|4.95|4.95|4.95|4.95|4.95|4.95|4.95|5.89|

Fig. 10.21 Joints restricted to lower floors[10.19]

10.5   Conclusion

In the design of any large concrete building it is essential in the preliminary stages to investigate the sensitivity of the structure to the various deformations and restraint forces which result from external load-independent influences, such as temperature changes and foundation movements, and internal effects such as shrinkage of the concrete. The resulting restraint forces and deformations can often be controlled by the choice of an appropriate structural system and structural design, and using an appropriate concrete mix with good curing. The design requirements of adequate strength and good serviceability can be achieved if the non-linear restraint forces are determined as part of the design process, with the inclusion of adequate minimum reinforcement in all regions of the structure.

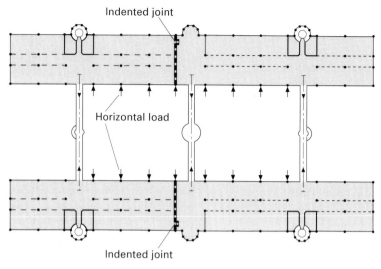

Fig. 10.22 Prefabricated building with joint which transmits shear force (Züblin Building)[10.20]

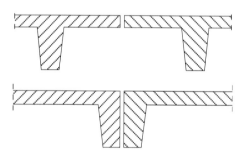

Fig. 10.23 Joints with complete separation

(a) Joint between girder and concrete hollow-core planks

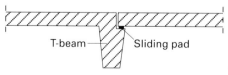

(b) Approach slab or T-beam

Fig. 10.24 Sliding joints

The decision to include joints should only be taken if the investigation shows that this is unavoidable; and as few joints as possible should then be employed. These should function as simply as possible, in order to have good durability. A well-conceived construction plan is needed to ensure the requirements of impermeability and aesthetics are met in the joint regions. The construction of the joints must be simple and economical.

References

10.1 Hock B, Schäfer K, Schlaich J 1986 Fugen und Aussteifungen in Stahlbetonskelettbauten. *DA f Stb* H 368 (Joints and stiffening regions in reinforced concrete skeletal construction)

10.2 Rybicki R 1972 *Schäden und Mängel an Baukonstruktionen* Werner-Verlag, Düsseldorf (Damage and defects in buildings)

10.3 Pfefferkorn W, Steinhilber H 1990 *Ausgedehnte fugenlose Stahlbetonbauten* Beton-Verlag, Düsseldorf (Extended, jointless construction in reinforced concrete)

10.4 Rybicki R 1977 *Faustformeln und Faustwerte für Konstruktionen im Hochbau – Teil 1: Geschossbauten* Werner-Verlag, Düsseldorf (Approximate formulae and values for design of high-rise buildings)

10.5 Katzenbach R 1993 Zur technisch-wirtschaftlichen Bedeutung der kombinierten Pfahl-Plattengründung, dargestellt am Beispiel schwerer Hochhäuser, *Bautechnik* **70**(3): 161–170 (On the technical and economic importance of combined pile-slab footings, considered in relation to large tall buildings)

10.6 Buch G, Paul H 1979 Zum Bau des Verwaltungshochhauses der Dresdner Bank AG in Frankfurt *Beton- und Stahlbetonbau* H 6: 137–144 (The construction of the administration building of the Dresden Bank, AG, in Frankfurt)

10.7 Gravert F W, Adt H E 1980 Besonderheiten bei der Planung des Hochhaustragwerks der Hessischen Landesbank in Frankfurt. *Beton- und Stahlbetonbau* H 8: 181–186 (Special considerations in the planning of a tall building structure for the Hessische Landesbank in Frankfurt)

10.8 Schäfer H G 1972 Das Verwaltungsgebäude G1 im Olympischen Dorf, München. *Beton- und Stahlbetonbau* H 7: 145–153 (The administration building G1 in the Olympic village in Munich)

10.9 Brandt J, Rösel W, Schwern D, Stöffler J 1993 *Betonfertigteile im Skelett- und Hallenbau* Fachvereinigung Deutscher Betonfertigteilbau e.V., Bonn, pp 1–88 (Precast construction of skeletal buildings and halls)

10.10 Schlaich J 1976 Fugen im Hochbau – wann und wo? *Der Architekt* H 4: 161–167 (Joints in buildings – when and where?)

10.11 Schober H 1994 Zum Funktionsneubau des Katharinenhospitals Stuttgart. *Stahlbau* **63**(5): 129–133 (Functional new construction of the Katharinen Hospital in Stuttgart)

10.12 Pfefferkorn W 1974 Konstruktive Besonderheiten beim Geno-Haus in Stuttgart. *Beton- und Stahlbetonbau* H 8: 235–239 (Design considerations for the GENO Building in Stuttgart)

10.13 Falkner H 1983 Fugenlose und wasserundurchlässige Stahlbetonbauten ohne zusätzliche Abdichtung. *Vortrag Betontag 1983* Deutscher Betonverein, Wiesbaden (Jointless and waterproof construction in reinforced concrete without additional sealing)

10.14   Falkner H 1984 Fugenloser Stahlbetonbau. *Beton- und Stahlbetonbau* H 7: 183–188 (Jointless reinforced concrete construction)

10.15   Franz G, Schäfer K 1988 *Konstruktionslehre des Stahlbetons, Band II: Tragwerke, Teil A: Typische Tragwerke* Springer-Verlag, Berlin, Heidelberg, New York (Design of reinforced concrete)

10.16   Schober H 1990 Diagramme zur Mindestbewehrung bei überwiegender Zwangbeanspruchung. *Beton- und Stahlbetonbau* H 3: 57–62 (Diagrams for minimum reinforcement of members with imposed deformations)

10.17   Schneider K-H, Bender K 1974 Konstruktion des Verwaltungshochhauses der Commerzbank AG in Frankfurt/Main. *Beton- und Stahlbetonbau* H 12: 274–282 (Design of the administration building of the Commerz Bank AG in Frankfurt)

10.18   Sommer H 1978 Messungen, Berechnungen und Konstruktives bei der Gründung Frankfurter Hochhäuser. *Bauingenieur* **53**: 205–211 (Measurement, calculation and design for foundations of high-rise buildings in Frankfurt)

10.19   Boll K 1974 Anordnung von Dehnfugen bei tragenden Skeletten des Hochbaus. *Bautechnik* H 3: 94–98 (Arrangement of expansion joints in skeletal building construction)

10.20   Steinle A 1985 Das Züblin-Haus. Anordnung von Bewegungsfugen in Bauwerken und Bauteilen. *Fertigteilforum* T1TGL 22903, Sept, pp 3–10 (The precast Züblin Building; arrangement of movement joints in buildings and elements)

10.21   Ruth J 1993 *Werkstoffverhalten in Grenzflächenbereichen der Tragelemente von Bauwerken* Universität Stuttgart, Dissertation, Institut für Tragwerksentwurf und -konstruktion (Material behaviour in contact surface regions of load-bearing elements)

10.22   Ruth J 1991 Influence of contact surface problems on design practice. *IABSE Colloquium on Structural Concrete*, Stuttgart

Index